APPLIED MECHATRONICS AND MECHANICS

System Integration and Design

APPLIED MECHATRONICS AND MECHANICS

System Integration and Design

Edited by

Satya Bir Singh, PhD
Prabhat Ranjan, PhD
A. K. Haghi, PhD

First edition published 2021

Apple Academic Press Inc.
1265 Goldenrod Circle, NE,
Palm Bay, FL 32905 USA

4164 Lakeshore Road, Burlington,
ON, L7L 1A4 Canada

CRC Press
6000 Broken Sound Parkway NW,
Suite 300, Boca Raton, FL 33487-2742 USA

2 Park Square, Milton Park,
Abingdon, Oxon, OX14 4RN UK

First issued in paperback 2021

© 2021 Apple Academic Press, Inc.

Apple Academic Press exclusively co-publishes with CRC Press, an imprint of Taylor & Francis Group, LLC

Library and Archives Canada Cataloguing in Publication

Title: Applied mechatronics and mechanics : system integration and design / edited by Satya Bir Singh, PhD, Prabhat Ranjan, PhD, A.K. Haghi, PhD.
Names: Singh, Satya Bir, editor. | Ranjan, Prabhat, (Mechatronics professor), editor. | Haghi, A. K., editor.
Description: Includes bibliographical references and index.
Identifiers: Canadiana (print) 20200305255 | Canadiana (ebook) 20200305433 | ISBN 9781771888899 (hardcover) | ISBN 9781003019060 (PDF)
Subjects: LCSH: Mechatronics. | LCSH: Mechanics, Applied.
Classification: LCC TJ163.12 .A67 2021 | DDC 621—dc23

Library of Congress Cataloging-in-Publication Data

CIP data on file with US Library of Congress

ISBN: 978-1-77188-889-9 (hbk)
ISBN: 978-1-77463-916-0 (pbk)
ISBN: 978-1-00301-906-0 (ebk)

About the Editors

Satya Bir Singh, PhD
Professor, Department of Mathematics, Punjabi University, Patiala, India

Satya Bir Singh, PhD, is a Professor of Mathematics at Punjabi University Patiala in India. Prior to this, he worked as an Assistant Professor in Mathematics at the Thapar Institute of Engineering and Technology, Patiala, India. He has published about 125 research papers in journals of national and international repute and has given invited talks at various conferences and workshops. He has also organized several national and international conferences. He has been a coordinator and principal investigator of several schemes funded by the Department of Science and Technology, Government of India, New Delhi; the University Grants Commission, Government of India, New Delhi; and the All India Council for Technical Education, Government of India, New Delhi. He has 21 years of teaching and research experience. His areas of interest include the mechanics of composite materials, optimization techniques, and numerical analysis. He is a life member of various learned bodies.

Prabhat Ranjan, PhD
Assistant Professor in the Department of Mechatronics Engineering at Manipal University Jaipur, India

Prabhat Ranjan, PhD, is an Assistant Professor in the Department of Mechatronics Engineering at Manipal University Jaipur, India. He is the author of the book *Basic Electronics* and editor of the book *Computational Chemistry Methodology in Structural Biology and Materials Sciences*. Dr. Ranjan has published more than 10 research papers in peer-reviewed journals of high repute and dozens of book chapters in high-end edited research books. He has received prestigious the President Award–Manipal University Jaipur, India, in 2015 for the development of the university; a Material Design Scholarship from Imperial College of London, UK, in 2014; a DAAD Fellowship in 2014; and CFCAM–France Award in 2015. Dr. Ranjan has received several grants and has also participated in national and international conferences and summer schools. He holds a Bachelor of Engineering in Electronics and Communication and a Master of Technology in Instrumentation Control System Engineering from the Manipal Academy of Higher Education, Manipal, India; as well as PhD in engineering from Manipal University Jaipur, India.

A. K. Haghi, PhD

Professor Emeritus of Engineering Sciences, Former Editor-in-Chief, International Journal of Chemoinformatics and Chemical Engineering and Polymers Research Journal; Member, Canadian Research and Development Center of Sciences and Culture

A. K. Haghi, PhD, is the author and editor of 165 books, as well as 1000 published papers in various journals and conference proceedings. Dr. Haghi has received several grants, consulted for a number of major corporations, and is a frequent speaker to national and international audiences. Since 1983, he served as professor at several universities. He is former Editor-in-Chief of the *International Journal of Chemoinformatics and Chemical Engineering and Polymers Research Journal* and is on the editorial boards of many international journals. He is also a member of the Canadian Research and Development Center of Sciences and Cultures (CRDCSC), Montreal, Quebec, Canada. He holds a BSc in urban and environmental engineering from the University of North Carolina (USA), an MSc in mechanical engineering from North Carolina A&T State University (USA), a DEA in applied mechanics, acoustics and materials from the Université de Technologie de Compiègne (France), and a PhD in engineering sciences from Université de Franche-Comté (France).

Contents

Contributors

Purnank Bhatt
Assistant Professor, Mechanical Engineering Department,
G. H. Patel College of Engineering & Technology, Vallabh Vidyanagar, Gujarat, India

Vijay Chaudhari
Professor, Mechanical Engineering Department, C. S. Patel Institute of Technology,
Charusat University, Changa, Gujarat, India

Himanshu Chaudhary
Mechanical Engineering Department, Malaviya National Institute of Technology, Jaipur,
Rajasthan, India

Ram Dayal
Department of Mechanical Engineering, Malaviya National Institute of Technology, Jaipur,
Rajasthan, India

Kumar Gaurav
Department of Mechatronics Engineering, Manipal University Jaipur, Dehmi Kalan, Jaipur,
Rajasthan, India, E-mails: kumar.gaurav@jaipur.manipal.edu; kumarmuj@gmail.com

Vandana Gupta
Department of Mathematics, Dashmesh Khalsa College, Zirakpur, Mohali, Punjab, India,
E-mail: vaggarwal2584@gmail.com

A. K. Haghi
Professor Emeritus, Canadian Research and Development Center of Sciences and Cultures,
Montreal, Canada

Tung-Yung Huang
Department of Mechanical Engineering, Southern Taiwan University of Science and Technology,
Tainan, Taiwan, E-mail: huangt@stust.edu.tw

Shih-Ying Hung
Department of Mechanical Engineering, Southern Taiwan University of Science and Technology,
Tainan, Taiwan

Anand Joshi
Professor, Mechatronics Engineering Department, G. H. Patel College of Engineering & Technology,
Vallabh Vidyanagar, Gujarat, India

Ajay Kumar
Department of Mechatronics Engineering, Manipal University Jaipur, Dehmi Kalan, Jaipur,
Rajasthan, India

Narayan Kumar
Department of Mechanical Engineering, IIT Jodhpur, Karwar, Jodhpur – 342037, Rajasthan, India

Parveen Lata
Associate Professor, Department of Basic and Applied Science, Punjabi University Patiala, Punjab, India

T. Muthuramalingam
Department of Mechatronics Engineering, SRM Institute of Science and Technology, Kattankulathur – 603203, Tamil Nadu, India, E-mail: muthu1060@gmail.com

Chitresh Nayak
Professor, Oriental Institute of Science and Technology, Bhopal, Madhya Pradesh, India

Vimal Kumar Pathak
Assistant Professor, Department of Mechanical Engineering, Manipal University Jaipur, Dehmi Kalan, Jaipur – 303007, Rajasthan, India

G. Pohit
Department of Mechanical Engineering, Jadavpur University, Kolkata – 700032, West Bengal, India, E-mail: gpohit@gmail.com

Prabhat Ranjan
Assistant Professor in the Department of Mechatronics Engineering at Manipal University Jaipur, India, E-mail: prabhat.ranjan@jaipur.manipal.edu

B. Ravindra
Department of Mechanical Engineering, IIT Jodhpur, Karwar, Jodhpur – 342037, Rajasthan, India, E-mail: ravib@iitj.ac.in

Shivdev Shahi
Research Scholar, Department of Mathematics, Punjabi University, India, E-mail: shivdevshahi93@gmail.com

Prem Singh
Mechanical Engineering Department, Swami Keshvanand Institute of Technology, Jaipur, Rajasthan, India, E-mail: premsingh001@gmail.com

Ramanpreet Singh
Department of Mechanical Engineering, Manipal University Jaipur, Dehmi Kalan, Jaipur, Rajasthan, India, E-mail: ramanpreet.singh@jaipur.manipal.edu

Satya Bir Singh
Professor, Department of Mathematics, Punjabi University, Patiala, Punjab, India, E-mail: sbsingh69@yahoo.com

Sukhveer Singh
Assistant Professor, Punjabi University APS Neighborhood Campus, Dehla Seehan, Sangrur, Punjab, India, E-mail: Sukhveer_17@pbi.ac.in

Mihir Solanki
Assistant Professor, Mechanical Engineering Department, G. H. Patel College of Engineering & Technology, VallabhVidyanagar, Gujarat, India

Deepak Rajendra Unune
Assistant Professor, Department of Mechanical Engineering, The LNM Institute of Technology, Jaipur, Rajasthan, India

Abbreviations

ADC	analog to digital converter
ANOVA	analysis of variance
BFO	bacterial foraging optimization
BMVSS	Bhagwan Mahaveer Viklang Sahayata Samiti
BWSAS	Bouc-Wen-based semi-active suspension system
COG	centroid of gravity
COTS	commercially off the shelf
CRONE	Commande Robuste d'Ordre Non-Entier
DAQ	data acquisition
DF	degree of freedom
ECU	engine control unit
EDM	electrical discharge machining
FFS	flexible force sensor
FGMs	functionally graded materials
FG-MWCNTs	functionally graded multi-walled carbon nanotubes
FSR	force-sensing resistors
GA	genetic algorithm
H	high
HAP	hydroxyapatite
HIL	hardware-in-loop
HSS	high-speed steel
HT	height
HTLPSO	hybrid-teaching learning particle swarm optimization
KAFO	knee-ankle-foot orthoses
L	low
LEER	lower extremity exoskeleton robot
M	medium
MCMC	Markov chain Monte Carlo
MGA	modified genetic algorithm
MOOP	multi-objective optimization problem
M-PSO	modified particle swarm optimization
MR	magnetorheological
NHTSA	National Highway Traffic Safety Administration

NSJAYA	nondominated sorting Jaya algorithm
OSL	odor source localization
PaS	passive suspension system
PGO	powered gait orthosis
PID	proportional-integral-derivative
PSO	particle swarm optimization
PTB	patella tendon bearing
RF	radio frequency
RMSE	root mean square error
SDE	source declaration error
SiCp	silicon carbide particles
SIL	software-in-loop
SL	stump length
SUBAR	Sogang University biomedical assistive robot
TPMS	tire pressure maintenance system
TPMS	tire pressure monitoring sensor
VA	virtual agents
VH	very high
VL	very low
WT	weight

Preface

Mechatronics is a core subject for engineers, combining elements of mechanical and electronic engineering into the development of computer-controlled mechanical devices. A mechatronics system integrates various technologies involving sensors, measurement systems, drives, actuation systems, microprocessor systems, and software engineering.

A universally accepted definition of the term "mechatronics" is the integration of several disciplines such as mechanics, electronics, electrical, computer, control, and software engineering using microelectronics to control mechanical devices.

Applied mechatronics and mechanics is the integration of mechanical engineering, electronic engineering, control, and computer engineering. Synergistic collaboration among these fields of science involves a high potential for accomplishments and achievements now accessible to a wide variety of engineers.

This research-oriented book presents a clear and comprehensive introduction to the area and helps the readers to comprehend and design mechatronic systems by providing a frame of understanding to develop a truly interdisciplinary and integrated approach to engineering.

The design of mechatronics products is a challenge for design engineers. Therefore, the mechatronic process is a cross-disciplinary design process that can be properly applied if, and only if, the specialists from all pertinent disciplines work together from a very early stage in the design process.

Applied Mechatronics and Mechanics is the most comprehensive research-oriented book available for both mechanical and electrical engineering students and will enable them to engage fully with all stages of mechatronic system design. It offers broader and more integrated coverage than other research-oriented books in the field with case studies.

This volume is a source of the latest research and technical notes in mechatronics. This book is useful for students, researchers, and all readers interested in this topic.

This volume focuses on application considerations and relevant practical issues that arise in the selection and design of mechatronics components and systems as well.

It provides practicing mechanical/mechatronics engineers and designers, researchers, graduate, and postgraduate students with knowledge of mechanics focused directly on advanced applications.

Chapter 1 presents the balancing procedure of the mechanism using a multi-objective Jaya algorithm. The mechanisms can be balanced by optimizing the inertial properties of each link. The inertial properties of each rigid link are represented by the dynamic equivalent system of point masses. Thus, the multi-objective optimization problem (MOOP) with the minimization of shaking forces and shaking moments is formulated by considering the point mass parameters as the design variables. The formulated optimization problem is solved by a posteriori approach-based algorithm as a multi-objective Jaya algorithm (MOJAYA) under suitable design constraints. This algorithm uses the concepts of crowding distance and a nondominated sorting approach to find a Pareto set of optimal solutions. The efficiency of the proposed approach is investigated through a four-bar planar standard mechanism taken from literature. It is established that the multi-objective Jaya algorithm is computationally more efficient than NSGA-II. The designer can choose any solution from the set and balance any real planar mechanisms.

In Chapter 2, different types of suspension systems are discussed. The advantage of a semi-active suspension system is highlighted. It is shown that MR damper is an important part of automotive suspension systems and, hence, plays a major role in controlling the dynamic nature of the vehicle. The effectiveness of the MR damper was studied. Initially, a quarter car model with a passive suspension system (PaS) was designed. Another quarter car model was developed with an MR damper as a member of the semi-active suspension system. The Bouc-Wen hysteresis model predicted the behavior of the MR damper. The final quarter car model was developed with an MR damper based on the Bouc-Wen model and controlled by a fuzzy-logic controller. These three models were analyzed by considering the step road profile. Altering the values of the scalar gains to produce better results optimized the fuzzy-logic controller.

In Chapter 3, the importance of enhancing the performances of conventional manufacturing processes is being increased. It is possible only where integrating the mechatronics approaches in manufacturing processes. The adaptation of control algorithms and modification of existing electronics circuits in the manufacturing processes can considerably enhance the performances of the manufacturing technology. The efficiency of the unconventional machining processes can be effectively increased by the mechatronics engineering concepts. In the present study, an endeavor has been made to analyze the various literature related to the adaptation of mechatronics knowledge in manufacturing technology. The responses of the processes with and without mechatronics approaches have been compared and the merits of those systems over the existing systems have been analyzed. The influences of the various control strategies have also been analyzed and compared. It has been observed that the still better research works can be made to increase the efficacy the manufacturing technology.

Chapter 4 applies the nature-inspired optimization-based approach for odor source localization (OSL) problem in an indoor environment with mobile agents. An optimization problem is formulated to identify the region of maximum odor concentration. The optimization problem is solved using a new hybrid-teaching learning particle swarm optimization (HTLPSO) algorithm. To demonstrate the effectiveness of the adopted approach, experiments are conducted in a simulated environment generated through the Gaussian plume model. A various set of mobile agents in a range of {3–15} are used to find and detect the source of odor with high accuracy and concentration, respectively. Furthermore, it is found that less number of mobile agents is required for fast convergence to detect the odor source.

Tire pressure monitoring systems as a case study in automotive mechatronics presented in Chapter 5. Chapter 6 is a retrospective assessment of elastic-plastic and creep deformation behavior in structural components made of discs, cylinders, and spherical/cylindrical shells. Unlike elastic solids in which the state of strain depends only on the final state of stress, the deformation that occurs in a plastic solid is determined by the complete history of loading. The plasticity problem is, therefore, essentially incremental in nature, the final distortion of the solid being obtained as the sum total of the incremental distortions following the strain path. The deformation character has been considered to be linear in classical theory which is not accurate. The non-linear character was studied by B. R. Seth considering the transition surface function and generalized strain measure to determine elastic-plastic and creep state. A number of elastic-plastic and creep problems pertaining to various structural components, made of materials exhibiting different kinds of isotropy and anisotropy have been solved using this transition theory and are reviewed in this chapter. It is sufficient to say that the transition functions, which define the non-linearity of elastic-plastic and creep transition, are more accurate when compared to the classical theory.

In Chapter 7, elastic-plastic transition stresses in Zirconia-based crowns of ceramic dental implants have been calculated analytically. The crown of the implant is modeled in the form of a shell which exhibits transversely isotropic macro-structural symmetry. The transition theory of B. R. Seth has been used to model the elastic-plastic state of stresses. The shell so modeled is subjected to external pressure to analyze the state of axial compression. The results for Zirconia-based implants are compared with a titanium-based dental implant. The elastic stiffness constants for these are taken from the available literatures which have been obtained using ultrasonic resonance spectroscopy, a non-destructive technique of obtaining the stiffness constants. The radial and circumferential stresses are obtained for radius ratios, which can handle any type of dataset for thicknesses of crowns.

Recent developments in the theory of nonlocality in elastic and thermoelastic mediums; presented in Chapter 8.

In Chapter 9, a mathematical model is developed to investigate the creep response in rotating composite disc with hyperbolic thickness for isotropic material. The disc is supposed to be made of the composite containing silicon carbide particles (SiCp) embedded in a matrix of pure aluminum. The model is used to investigate the effect thermal gradients on the stresses and strain rates of the isotropic disc by Sherby's creep law and von Mises criterion of yielding. It is concluded that thermal gradients introduce a significant change in the strain rates although its effect on the resulting stress distribution is relatively small in a composite rotating disc.

The aim of Chapter 10 is to find an adequate rule of tuning fractional-order PID controllers for dynamic systems. PID controllers of integer order have been successfully employed in versatile applications, especially with the aid of available mature tuning methods such as the Ziegler-Nichols method. In recent years, fractional-order PID controllers tuned with the Ziegler-Nichols type method were proposed to deal with fractional-order systems. It is intriguing for the flexibility of the adoption of fractional order which is free from the limit of integer order. This article proposes a Ziegler-Nichols type tuning method for the fractional order PID controller applied in an underactuated system, and studies the effect of fraction order. The simulation results demonstrate that the proposed method works effectively.

Chapter 11 presents a methodology for evaluating the role of amputee's physical parameters viz. height, weight, and stump length (SL) on the pressure generated at the prosthetic limb/socket interface using the regression technique and fuzzy-logic-based model. Intra-socket pressures can cause the tissue trauma or discomfort to amputees wearing the prosthetic devices. The intra-socket pressure values at lateral tibia, gastrocnemius, patella tendon bearing (PTB), kick point, medial tibia, medial gastrocnemius, popliteal depression, and lateral gastrocnemius on the transtibial residual limb have been collected for three different conditions viz. walking, full load and half load for ten patients. An experimental setup is developed for force investigation of the lower-limb socket using the Flexi-Force sensor. The experimental trials are performed and further experimental data are used to establish the Mamdani fuzzy-logic model to predict the effective pressure at considered specific regions as the response parameters. Mathematical models for pressure at three loading conditions have been developed using regression analysis using which the effective pressure at considered specific regions can be correlated with physical parameters. The models suggested that the weight and followed by stump length of amputees are a strong predictor of pressure at the socket. The confirmation experiment results reveal that the fuzzy model shows good agreement of 98.53% with the experimentally measured value that proves that the established fuzzy model consequently can be used for predicting the effective pressure at the socket interface for different points. The developed methodology will assist the amputee-specific socket design ensuring comfortable socket fitting.

Experimental investigation and optimization of process parameters in oblique machining process for hard to cut materials coated inserts presented in Chapter 12.

Chapter 13 presents the mechanical design of a slider-crank mechanism for a knee joint orthotic device. The device is portable for stroke patients and helps in gait training at home, clinics, and hospitals. The knee joint mechanism is optimized by considering gait biomechanics for making it potable. An optimization problem is formulated by considering the geometrical parameters of slider-crank linkage with an objective to reduce the required peak force by an actuator while walking on a leveled ground. Based on the optimization problem, constraints are posed to ensure the range of motion as in normal human gait. The optimization problem is solved using the Jaya algorithm. It is observed that there is a significant reduction in the peak force required by the actuator.

CHAPTER 1

Dynamic Balancing of Planar Mechanisms Using Nondominated Sorting Jaya Algorithm

PREM SINGH,[1] RAMANPREET SINGH,[2] and HIMANSHU CHAUDHARY[3]

[1]Mechanical Engineering Department, Swami Keshvanand Institute of Technology, Jaipur, Rajasthan, India, E-mail: premsingh001@gmail.com

[2]Department of Mechanical Engineering, Manipal University Jaipur, Rajasthan, India

[3]Mechanical Engineering Department, Malaviya National Institute of Technology, Jaipur, Rajasthan, India

ABSTRACT

This chapter presents the balancing procedure of the mechanism using a nondominated sorting Jaya algorithm (NSJAYA). Planar mechanisms can be balanced by optimizing each moving link's inertial properties. These properties are derived by the point-mass system. Thus, the multi-objective optimization problem (MOOP) is posed to minimize the unbalance force and moment in which the point mass parameters are treated as the design variables. The formulated optimization problem is solved by a posteriori approach-based algorithm as a NSJAYA under suitable design constraints. This algorithm uses the concept of crowding distance and a nondominated sorting approach to find a Pareto set of optimal solutions. The proposed method is tested through a four-bar planar standard mechanism taken from literature. It is established that the NSJAYA is computationally more efficient than NSGA-II. The designer can choose any solution from the set and balance any real planar mechanisms.

1.1   INTRODUCTION

Typically, the unbalanced mechanism develops the forces and moments on the bearings which are also known as shaking forces and moments. They increase the vibration, driving torque, fatigues, etc., in the mechanism. Inertia, the center of mass, and mass of moving link define the input torque, the forces, and moments [1]. Recently, the balancing of these forces and moments is a challenging task. Therefore, many techniques have been applied to reduce the forces and moments using various principles such as counterweights [2] and the redistribution of the masses [3] are used to minimize the forces. But, this force balancing procedure generally increases moments and input torque of the mechanism [4]. Moreover, disk or inertia counterweights [5, 6], moment balancing idler loops [7], and a replicate mechanism [8] are used to minimize the moments. These approaches of balancing enhance the mass and complicacy of the mechanism [4].

In contrast, the optimization procedure has also been used by researchers to balance the mechanisms. Typically, two objectives, namely, shaking force and shaking moment are chosen. Thus, the optimization problem with multi-objectives is posed to minimize these objectives [9]. The formulated problem can be solved using two approaches as *a priori* approach and a posteriori approach [10]. The first approach obtains a single objective problem from the multi-objective problem using an appropriate weight for each objective function. This approach gives a unique optimal design in each simulation run. Therefore, multiple optimal solutions are obtained with a different combination of weights. Moreover, the optimum results obtained by this approach depend upon the weights assigned to each objective function. Therefore, the designer must know the importance of each objective function while assigning the weights to the objective functions that can be difficult for an uncertain scenario. In addition, a *posteriori* approach eliminates the drawbacks of *a priori* approach. In this approach, the weights are not assigned to the objectives before the start of the algorithm. It provides Pareto optimal solutions in the single run of the algorithm. The user can choose appropriate solutions based on the importance of objective functions from the Pareto optimal set of solutions [11]. This makes the posteriori approach computationally more efficient in comparison to the priori approach. Therefore, it is applied for the balancing problems of the planar mechanism.

The conventional optimization techniques like gradient search methods based on *a priori* approach can be applied to balance the shaking force and moment [12, 13]. However, they need an initial start point to find the optimal solution [14]. Thus, these algorithms give the local solution near to start point. Moreover, nature-inspired optimization techniques like the genetic algorithm (GA) [15, 16] and particle swarm optimization (PSO) [17], and a hybrid of two optimization algorithms [18] can be applied to balance the mechanisms. But, these optimization techniques require specific parameters for their convergence.

No relevant research has been published in which a posteriori approach based algorithm is applied to balance the mechanisms. In this study, a posteriori approach based algorithm as a nondominated sorting Jaya algorithm (NSJAYA) is applied to balance the planar mechanism. The shaking forces and shaking moments are defined using the concept of the dynamically equivalent point mass system [19]. This system represents the inertial properties of each moving link. In order to balance the mechanism, the optimization problem with multi objectives as shaking force and shaking moment are stated by considering the parameters of point masses as the design variables. The efficiency of the proposed method is validated through a standard four-bar planar mechanism taken from the literature. It is found that the NSJAYA algorithm is more efficient in computation than NSGA-II used in this study. The Pareto optimal set for the balancing problem is calculated and outlined in this chapter. The designer can balance any real planar mechanisms using any solution from the Pareto optimal set of solutions.

This chapter is structured as follows: Section 1.2 describes the dynamic analysis of the planar mechanism; Section 1.3 describes the optimization problem formulation using the concept of a point mass system; Section 1.4 presents the optimization algorithm. Results and discussion are outlined in Section 1.5; and finally, Section 1.6 outlines the conclusions.

1.2 DYNAMIC ANALYSIS OF THE PLANAR MECHANISM

This section presents the determination of the shaking forces and shaking moments developed in the four-bar planar mechanisms as shown in Figure 1.1. $\{o_1 xy\}$ and $\{o_i x_i y_i\}$ are the global and local coordinate systems, respectively. The reaction forces at joints are determined using Newton-Euler equations [20]. Then, these forces determine the shaking force and shaking moment. The shaking force is the summation of transmitted forces to fixed link in vector form [21] while the vector sum of driving torques and moments of the reaction forces about the fixed points is known as shaking moment [22].

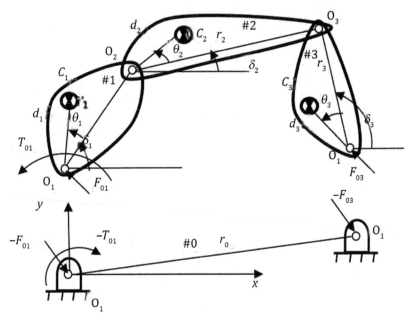

FIGURE 1.1   Representation of reaction forces and moments.

Thus, shaking forces and shaking moments are transferred to the fixed link '#0,' given as:

$$f_{sh} = -\left(F_{01} + F_{03}\right) \tag{1.1}$$

$$M_{sh} = -T_{01} - r_0 \times F_{03} \tag{1.2}$$

where F_{01} and F_{03} are 2-vector reaction forces of link #1 and link #3 at link #0, respectively, while T_{01} is driving torque about joint o_1. r_4 is the vector from o_1 to o_4.

1.3   OPTIMIZATION PROBLEM FORMULATION

This section formulates the optimization problem with multi objectives as the shaking force and shaking moment using a point-mass system approach. The mechanism shown in Figure 1.1 has three moving links. Each moving link defined as in Figure 1.2(a) is systematically converted into the system of three-point masses as shown in Figure 1.2(b). The point-mass parameters are treated as the design variables. These parameters describe the shaking force and moment which are reported in Eqs. (1.1)–(1.2). Generally, three parameters need to specify each point mass while each link needs nine parameters to represent it. The vector form of *i*-th link's design variables is represented as:

$$x_i = \left[m_{i1} \theta_{i1} a_{i1} m_{i2} \theta_{i2} a_{i2} m_{i3} \theta_{i3} a_{i3} \right]^T \tag{1.3}$$

where m_{ij} is the *j*-th point mass for the *i*-th link, a_{ij} and θ_{ij} are the corresponding length and angular position of m_{ij}.

Thus, a total of 27 point mass parameters defines the mechanism, given as:

$$x = [x_1^T x_2^T x_3^T]^T \tag{1.4}$$

The multi-objective optimization problem (MOOP) is finally stated under the appropriate constraints by taking the R.M.S values of shaking force, $F_{sh,rms}$ and the shaking moment, $M_{sh,rms}$ as:

Minimize $\qquad\qquad f_1(x) = F_{sh,rms} \qquad\qquad\qquad (1.5)$

Minimize $\qquad\qquad f_2(x) = M_{sh,rms} \qquad\qquad\qquad (1.6)$

Subjected to:

$$\left.\begin{aligned}
g_{1i}(x) &= m_{i,min} - \sum_j \text{mij} < 0 \\
g_{2i}(x) &= \sum_j \text{mij} - m_{i,max} < 0 \\
g_{3i}(x) &= d_i - r_i < 0 \\
g_{4i}(x) &= m_i d_i^2 - I_i \\
LB_r &\le x_r \le UB_r r = 1, \dots N
\end{aligned}\right\} \; for\, i, j = 1, \dots, 3 \tag{1.7}$$

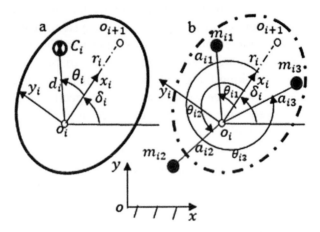

FIGURE 1.2   (a) *i*-th moving rigid link and (b) point-mass system for *i*-th moving link.

The design constraints depend on the permissible values of the parameters of the link [20]. While the strength of links defines the maximum and minimum masses of the link, similarly, the allowable values of the moment of inertias determine the limits on parameters, a_{ij} [23]. UB_r and LB_r are the upper and lower limits of the *r*th design variable. While the number of design variables is denoted by N.

Moreover, an unconstrained optimization problem is obtained from the constrained multi-objective problem defined in Eqs. (1.5)–(1.7) using penalty concept [24]. A significant penalty value is utilized to the objective function for the violation of each constraint [25]. Thus, the optimal solution is determined that satisfies all the constraints using a suitable optimization algorithm. Then, the optimization problem is stated in term of penalty and objective function as:

Minimize $f_1(x) = f_{sh,rms} + pn \times \sum_{c=1}^{n_c} \max\left(0, g_i(x)\right)$ (1.8)

Minimize $f_2(x) = M_{sh,rms} + pn \times \sum_{c=1}^{n_c} \max\left(0, g_i(x)\right)$ (1.9) where n_c is the number of constraints, while *pm* is the penalty value of the order of 10^4 which assigned to the objective function for constraint violation [24]. Finally, the formulation of the MOOP is given as:

$$\text{Minimize } Z(x) = \left[f_1(x) f_2(x) \right]^T \tag{1.10}$$

$$LB_r \le x_r \le UB_r r = 1, \ldots N \tag{1.11}$$

1.4 OPTIMIZATION ALGORITHM

The formulated problem as described in the previous section can be solved by posteriori based algorithms like GA [26], MPSO [27], PSO [28], NSGA-II [29], etc. Although these algorithms give a global solution, it is no assurance that the solution is optimal [24]. Moreover, these algorithms need algorithmic control parameters, which affect their performance [30]. However, the NSJAYA algorithm is also the posteriori based algorithm which does not require an algorithm-specific parameter. The solutions obtained using this algorithm are improved in a similar way as in the Jaya algorithm, which proposed by Rao in 2016 [30]. Moreover, the NSJAYA algorithm uses the crowding distance computation mechanism and nondominated sorting approach to solve the MOOP as given in Ref. [11]. Also, the determination of best and worst solution in multi-objectives is difficult due to the multiple opposing objectives.

Therefore, the NSJAYA algorithm is developed to find the best and worst solutions. This algorithm uses the ranking approach of solutions based on crowding distance value and the non-dominance concept to find these solutions. In starting, initial solutions "p" are randomly created. Then, initial solutions are arranged in ascending order with rank using the non-dominance approach [30]. The best and worse solutions are selected based on higher and lower rank, respectively. If solutions have the same rank, then crowding distance value is calculated for these solutions. Then, the worse solution (low crowding distance value) and the best solution (large crowding distance value) are identified in each iteration. The solutions are modified according to the general equation of the Jaya algorithm [11]. Then, the population-based on updated solutions is merged with the initial solutions, and 2p populations are generated. The ranks of these merged solutions are arranged again. Further, the value of crowding distance for every solution is determined. The best solutions are chosen according to the new rank and value of crowding distance. The higher rank and more crowding distance value solutions are chosen as the best solution than other solutions [31]. The rank of solutions based on the dominant approach, the computation of crowding distance, and the operator of crowding comparison are detailed in Refs. [11, 29, 31].

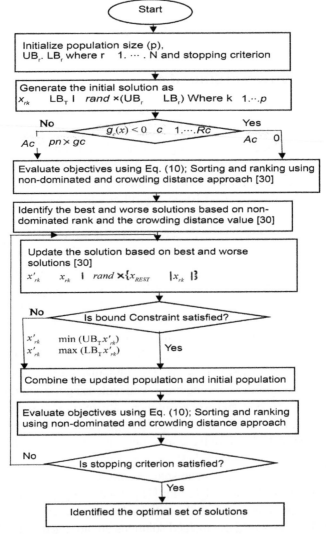

FIGURE 1.3　A flow chart for the nondominated sorting Jaya algorithm (NSJAYA).

This algorithm is terminated based on the number of function evaluations and generations. The optimization procedure of the algorithm is described in Figure 1.3. Further, this algorithm decreases the computational time than the other multi-objective optimization techniques. Moreover, the first time it is applied to the dynamic balancing of the planar mechanism in this study.

1.5 RESULTS AND DISCUSSIONS

This section describes the effectiveness of the proposed method by the balancing of the four-bar mechanism taken from the literature [17]. In this research, a three-point mass system is used for each moving link of the mechanism as shown in Figure 1.2(b). The five parameters of each link are assigned as [20]:

$$\Theta_{i1} = 0; \; \Theta_{i2} = 120°; \; \Theta_{i3} = 240°, \text{ and } a_{i1} = a_{i2} = a_{i3}.$$

Hence, m_{i1}, m_{i2}, m_{i3}, and a_{i1} other parameters of each link are taken as the design variables and the upper and lower bound of, m_i, d_i and a_i are taken as, $m_i^o \leq m_i \leq 2m_i^o \;\; 0 \leq d_i \leq r_i$, and $0.5a_{i1}^o \leq a_{i1} \leq 1.5a_{i1}^o$, respectively.

A planar four-bar mechanism [23] is balanced using the proposed approach. Link #1 rotates at a constant speed of 100 rad/s [23]. The link parameters of the original mechanism are presented in Table 1.1.

TABLE 1.1 Link Parameters of the Original Mechanism [23]

Link i	r_i (m)	m_i (kg)	I^G_i (kg-m^2)	Mass Center Location	
				d_i(m)	Θ_i(deg)
#1	0.1	0.392	0.0004	0.05	0
#2	0.4	1.570	0.213	0.2	0
#3	0.3	1.177	0.0091	0.15	0
#0	0.3	–	–	–	–

The optimization problem for multi objectives defined in Eqs. (10–11) is solved using the NSJAYA algorithm. MATLAB code is developed for this algorithm. To compare the computation efficiency of the algorithm, the same optimization problem is handled by NSGA-II. The detailed procedure of this algorithm is described in Ref. [29]. NSGA-II and NSJAYA algorithms take

200 iterations while NSGA-II and NSJAYA algorithms consider the population size of 150 and 100, respectively. Moreover, NSGA-II needs the tuning of specific parameters, whereas the NSJAYA algorithm is a parameterless algorithm. Hence, the default parameters are considered for NSGA-II. The function evaluations of 30,000 and 20,000 are taken by NSGA-II and NSJAYA algorithms to find the nondominated set of solutions, respectively. These algorithms provide the nondominated set of solutions as presented in Figure 1.4. The figure shows that the NSJAYA algorithm gives better results than the NSGA-II algorithm. It requires 33.33% lesser function evaluations than the NSGA-II. Thus, the NSJAYA algorithm is computationally more efficient compared to NSGA-II for the balancing of the mechanism. The nondominated solutions are obtained using the NSJAYA algorithm, and ten solutions out of them are reported in Table 1.2. Further, the designer can select any solution out of them based on the importance of the objective functions. Table 1.2 shows that the nondominated solutions with higher rank are in descending order of the crowding distance. Solutions 1 and 2 have the same crowding distance and considered as best solutions than other solutions. These solutions are likened with those of the original mechanism and solutions obtained by *a priori* approach based on algorithms like TLBO and GA [23] as shown in Table 1.3. The variations of the shaking force and the shaking moment for one cycle of the crank are shown in Figure 1.5.

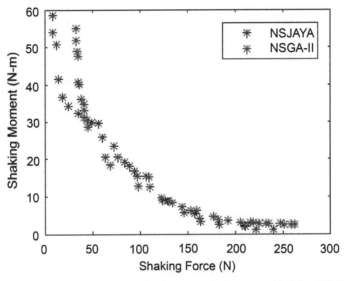

FIGURE 1.4 Comparison of non-dominance solutions provided by NSJAYA and NSGA-II.

TABLE 1.2 The Nondominated Solutions Obtained Using NSJAYA

Sl. No.	Shaking Force (N)	Shaking Moment (N-m)	Crowding Distance
1.	8.280534	58.47477	∞
2.	239.4591	1.126149	∞
3.	14.48232	41.36432	0.01108
4.	13.15304	50.6761	0.010124
5.	182.8893	2.608544	0.008765
6.	35.20479	32.31548	0.007557
7.	60.41955	26.00544	0.007477
8.	18.62965	36.61707	0.006981
9.	109.8948	12.70748	0.006883
10.	72.73956	23.60199	0.006652

Table 1.3 shows that *solution 1* minimizes the R.M.S values of the shaking force and the shaking moment up to 99.65% and 86.24%, respectively while *solution 2* minimizes 89.94% and 99.73% in the R.M.S values of the force and moment, respectively, compared to the original mechanism values. It shows that the NSJAYA algorithm provides better results than those of GA and TLBO.

TABLE 1.3 The R.M.S Values of Shaking Force and Shaking Moment

	Shaking Force (N)	Shaking Moment (N-m)
Original Mechanism [23]	2.3807e+03	424.9080
GA [23] (w_1= 0.5; w_2 =0.5)	917.28 (–60.65%)	144.95 (–65.52%)
TLBO [23] (w_1= 0.5; w_2 =0.5)	811.77 (–65.30)	112.161 (–73.36%)
Solution 1	8.280534 (–99.65%)	58.47477 (–86.24%)
Solution 2	239.4591 (–89.94%)	1.126149 (–99.73%)

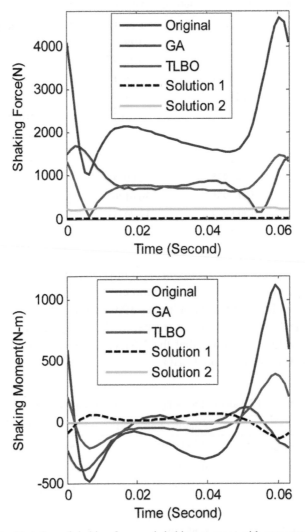

FIGURE 1.5   Variation of shaking force and shaking moment with respect to time.

1.6   CONCLUSIONS

The optimization procedure of multi-objectives to balance the planar mechanism is described in this chapter. The optimization problem with multi objectives as shaking forces and shaking moments is stated by applying the dynamically equivalent point-mass system. A posteriori approach based algorithm as NSJAYA is proposed for a four-bar planar mechanism balancing taken from Ref. [23]. This algorithm takes 33.33% fewer function evaluations for finding the optimal non-dominance solutions than NSGA-II. Hence, it is computationally more efficient compared to NSGA-II.

Solutions 1 and 2 correspond to a higher rank, and larger crowding distance obtained using NSJAYA are likened with the solutions generated by *a priori* approach based on algorithms like TLBO and GA. Solutions 1 and 2 reduce the R.M.S values of the shaking force and shaking moment by (99.65%, 86.24%) and (89.94%, 99.73%), respectively. Thus, the NSJAYA algorithm provides better results than those of GA and TLBO.

The proposed approach helps the designer for choosing the more alternatives based on the importance of objectives.

KEYWORDS

- **balancing**
- **four-bar mechanism**
- **Jaya algorithm**
- **multi-objective optimization problem**
- **Pareto set**
- **point mass parameters**

REFERENCES

1. Chaudhary, H., & Saha, S. K., (2007). Dynamic performance improvement of a carpet scrapping machine. *J. Sci. Ind. Res. (India), 66*(12), 1002–1010.
2. Walker, M. J., & Oldham, K., (1978). A general theory of force balancing using counterweights. *Mech. Mach. Theory, 13*(2), 175–185.
3. Kochev, I. S., (1987). General method for full force balancing of spatial and planar linkages by internal mass redistribution. *Mech. Mach. Theory, 22*(4), 333–341.
4. Chaudhary, H., & Subir, K. S., (2008). *Dynamics and Balancing of Multibody Systems* (37th edn.). Springer Science & Business Media.
5. Berkof, R. S., (1973). Complete force and moment balancing of inline four-bar linkages. *Mech. Mach. Theory, 8*(3), 397–410.
6. Esat, I., & Bahai, H., (1999). Theory of complete force and moment balancing of planer linkage mechanisms. *Mech. Mach. Theory, 34*(6), 903–922.

7. Bagci, C., (1979). Shaking force balancing of planar linkages with force transmission irregularities using balancing idler loops. *Mech. Mach. Theory, 14*(4), 267–284.

8. Arakelian, V. H., & Smith, M. R., (2005). Shaking force and shaking moment balancing of mechanisms: A historical review with new examples. *J. Mech. Des., 127*(2), 334–339.

9. Chaudhary, K., & Chaudhary, H., (2015). Shape optimization of dynamically balanced planar four-bar mechanism. *Procedia Computer Science*, pp. 519–526.

10. Rao, R. V., Rai, D. P., & Balic, J., (2019). Multi-objective optimization of abrasive water jet machining process using Jaya algorithm and PROMETHEE method. *J. Intell. Manuf., 30*(5), 2101–2127.

11. Rao, R. V., Rai, D. P., & Balic, J., (2018). Multi-objective optimization of machining and micro-machining processes using nondominated sorting teaching-learning-based optimization algorithm. *J. Intell. Manuf., 29*(8), 1715–1737.

12. Chaudhary, H., & Saha, S. K., (2007). Balancing of four-bar linkages using maximum recursive dynamic algorithm. *Mech. Mach. Theory, 42*(2), 216–232.

13. Li, Z., (1998). Sensitivity and robustness of mechanism balancing. *Mech. Mach. Theory, 33*(7), 1045–1054.

14. Chaudhary, K., & Chaudhary, H., (2014). Optimum balancing of slider-crank mechanism using equimomental system of point-masses. *Procedia Technol., 14*, 35–42.

15. Erkaya, S., (2013). Investigation of balancing problem for a planar mechanism using genetic algorithm. *J. Mech. Sci. Technol., 27*(7), 2153–2160.

16. Chaudhary, K., & Chaudhary, H., (2014). Dynamic balancing of planar mechanisms using genetic algorithm. *J. Mech. Sci. Technol., 28*(10), 4213–4220.

17. Farmani, M. R., Jaamialahmadi, A., & Babaie, M., (2011). Multi-objective optimization for force and moment balance of a four-bar linkage using evolutionary algorithms. *J. Mech. Sci. Technol., 25*(12), 2971–2977.

18. Singh, R., Chaudhary, H., & Singh, A. K., (2017). A new hybrid teaching-learning particle swarm optimization algorithm for synthesis of linkages to generate path. *Sadhana-Acad. Proc. Eng. Sci., 42*(11), 1851–1870.

19. Hayat, A. A., & Saha, S. K., (2017). Identification of robot dynamic parameters based on equimomental systems. *Proceedings of the Advances in Robotics on—AIR 17* (pp. 1–6).

20. Chaudhary, H., & Saha, S. K., (2009). *Dynamics and Balancing of Multibody Systems*. Springer Verlag, Germany.

21. Chaudhary, H., & Saha, S. K., (2006). Equimomental system and its applications. *ASME 8ᵗʰ Bienn. Conf. Eng. Syst. Des. Anal. Am. Soc. Mech. Eng.*, pp. 33–42.

22. Zhang, D., & Wei, B., (2016). *Dynamic Balancing of Mechanisms and Synthesizing of Parallel Robots*. doi: 10.1007/978–3-319–17683–3.

23. Chaudhary, K., & Chaudhary, H., (2016). Optimal dynamic design of planar mechanisms using teaching-learning-based optimization algorithm. *Proc. Inst. Mech. Eng. Part C J. Mech. Eng. Sci., 203*(19), 3442–3456.

24. Singh, R., Chaudhary, H., & Singh, A. K., (2017). Defect-free optimal synthesis of crank-rocker linkage using nature-inspired optimization algorithms. *Mech. Mach. Theory, 116*, 105–122.

25. Singh, P., & Chaudhary, H., (2018). A modified Jaya algorithm for mixed-variable optimization problems. *J. Intell. Syst.*, 1–21.

26. Coello, C. A., (2000). An updated survey of GA-based multi-objective optimization techniques. *ACM Comput. Surv., 32*(2), 109–143.

27. Feng, L., Mao, Z. Z., & Yuan, P., (2012). An improved multi-objective particle swarm optimization algorithm and its application. *Kongzhi Yu Juece/Control Decis, 27*(9).

28. Parsopoulos, K. E., & Vrahatis, M. N., (2002). Particle swarm optimization method in multi-objective problems. *2002 ACM Symposium on Applied Computing (SAC 2002)*, pp. 603–607.

29. Deb, K., Pratap, A., Agarwal, S., & Meyarivan, T., (2002). A fast and elitist multi-objective genetic algorithm: NSGA-II. *IEEE Trans. Evol. Comput., 6*(2), 182–197.

30. Rao, R. V. R., (2019). *Jaya: An Advanced Optimization Algorithm and its Engineering Applications.* Springer International Publishing AG, part of Springer Nature, doi: doi. org/10.1007/978–3-319–78922–4.

31. Rao, R. V., Rai, D. P., & Balic, J., (2017). A multi-objective algorithm for optimization of modern machining processes. *Eng. Appl. Artif. Intell., 61*, 103–125.

CHAPTER 2

Vibration Control of a Car Suspension System Using a Magnetorheological Damper with a Fuzzy Logic Controller

G. POHIT

Department of Mechanical Engineering, Jadavpur University, Kolkata – 700032, West Bengal, India, E-mail: gpohit@gmail.com

2.1 INTRODUCTION

The suspension system of a car is responsible for the smooth ride and keeping control of the car. Specifically, the suspension system maximizes the friction between the tires and the road to provide steering stability and good handling. Additionally, suspension systems provide comfort to the passengers by limiting the impact of particular road irregularities to the car, as well as the passengers riding inside.

The suspension system is made up of several components, including the chassis, which holds the cab of the car. The springs along with the shock absorbers and struts support the vehicle's weight and absorb and reduce excess energy from road shocks. Finally, the anti-sway bar shifts the movement of the wheels and stabilizes the car [1].

2.1.1 HISTORY OF SUSPENSION SYSTEM

The earliest form of suspension would be the ox-drawn cart. It consisted of a platform swing placed on iron chains attached to the wheel frame of the carriage. By the 17th century, the iron chains were replaced with leather straps. This system remained the template for all other suspension systems till the 19th century. This is a very primitive form of a suspension and no modern automobile uses this kind of 'strap suspension' system.

The world's first automobile was built in 1885. It was developed to be a self-driven variation of the horse-drawn vehicles. However, the suspension system present in horse-drawn vehicles was designed for relatively low speeds and low loads. The same suspension system would not meet the high load and high-speed demands of an internal combustion engine used nowadays in an automobile.

The first workable spring-suspension required advanced metallurgical knowledge and skill, and only became possible with the advent of industrialization. The first patent for a vehicle with a spring suspension was registered by Obadiah Elliot. Each wheel had two leaf springs on both sides and the body of the vehicle was fixed directly on to the springs which were in turn attached to the axles. Within a decade, most British horse carriages followed a similar design; the light carriages used wooden springs to avoid taxation while the larger carriages used steel springs. The springs were usually made up of low carbon steel and were placed one on top of the other to give it a tandem arrangement.

Leaf springs have been in use since the early Egyptians. Leaf springs were used by ancient military engineers in the form of bows to power their siege engines-a device which was used to break down castle doors—with little success at first. The use of leaf springs in catapults was later refined and made to function properly many years later. Springs were not necessarily made of metal; a sturdy branch of a tree could also be used as a spring, such as a bow.

The British steel would not function effectively on America's rough roads. For this reason, the Abbot-Downing Company of Concord, New Hampshire re-introduced the leather strap suspension. The leather strap suspension produced a swinging motion instead of the previous bouncing vertical motion of the spring suspension.

In 1901, Mors of Paris was the first to fit an automobile with shock absorbers. The 'Mors Machine' had the advantage of being equipped with a "damped" suspension system, because of which Henry Fournier won the prestigious Paris-to-Berlin race on 20 June 1901. Fournier's superior time was 11 hours 46 minutes 10 seconds whereas the best competitor was Léonce Girardot in a Panhard with a time of 12 hours 15 minutes 40 seconds.

In 1906, coil springs first appeared in a production vehicle by the British Runabout Company, designed by British Motor Car Company. At present, coil springs are used in almost all cars. Leyland Motors introduced torsion bars in a suspension system in 1920. In 1922, the Lancia Lambda became the first automobile to have independent front suspension. This became more common in other cars from 1932 onwards. At present, most cars have independent suspension on all four wheels [2].

In the early 1930s, Monroe introduced the direct-acting telescopic shock absorber, which was derived from aircraft practice. This type of unit, based on the resistance due to forcing oil through orifices, has become the most popular damping device.

In 1985, dramatic advances in the application of electronic control to dampers started taking place resulting from the intense competition to improve the ride quality of the traditional passenger vehicle. In these systems, the damper adjustments were made as a function of a wide variety of sensed parameters such as body attitude, yaw rate, acceleration, disturbance frequencies, brake controls, steering input, etc. These types of suspension systems are known as active and semi-active suspension system whereas a conventional suspension system is known as a passive system. The control algorithms would vary from being relatively simple to highly complex [3].

2.1.2   FUNCTIONS OF SUSPENSION SYSTEMS

The main purpose of suspension systems are as follows:

- Providing vertical compliance, so the wheels can follow an uneven road, isolating the chassis from unwanted vibrations.
- Maintaining proper wheels geometry parameters such as camber, toe, and kingpin inclination angles.
- Reacting to different control forces produced by the tires-longitudinal (acceleration and braking) forces, lateral (cornering) forces, and braking and driving torques.
- Preventing chassis roll.
- Ensuring maximum contact between the tire and the road with minimal load variations [4].

2.1.3   TYPES OF SUSPENSION SYSTEMS

In terms of control, suspension systems can be divided into three broad categories:

1. Passive suspension;
2. Semi-active suspension; and
3. Active suspension.

A schematic diagram of the different suspension systems is shown in Figure 2.1.

The passive suspension system (PaS) is the traditional system which is employed in most passenger cars. The major components are the conventional hydraulic shock absorber, which is essentially a damper, and the coil spring as shown in Figure 2.1(a). The spring absorbs the vibrations coming from the road and the damper dissipates it. The movement of the vehicle greatly depends on the nature of the road. From the viewpoint of passengers, the ride is fairly comfortable and from the driver's point of view, ride handling is rather robust.

Active suspension allows movement of the wheels relative to the chassis through an on-board system. Active suspension is available in luxury cars such as BMW, Audi, Citroën, Mercedes-Benz, etc. There is no spring and damper system present, as shown in Figure 2.1(c). In this system, numerous sensors are present which provide information about the road condition, wheel rotation speed, ground clearance, engine vibrations, transmission vibrations, steering geometry, pedestrians, and adjacent vehicles. This information is then sent to the engine control unit (ECU). The ECU contains multiple processors and a variety of algorithms for different systems. The ECU processes the information and sends the appropriate voltage signal to the active suspension system. The active suspension system comprises many actuators. The voltage signal from the ECU produces movement of the chassis through the actuators. Actuators can be solenoid actuators, motorized actuators, stepper motors, synchronous motors, or thermal actuators. Active suspension systems provide a great deal of comfort to the passengers; however, they come up short in terms of vehicle handling.

Ride handling and ride comfort are two conflicting properties of any vehicle system. To improve one, the other parameter has to be sacrificed. Balancing these two quantities is a laborious task. While active and PaS are located on opposite ends of the spectrum, semi-active suspension follows a middle path and provides an optimal solution to the handling-comfort paradox.

Semi-active suspensions provide better ride comfort than passive suspensions and at the same time provide better ride handling than active suspensions. In the event the ECU malfunctions or ceases to functions or the wiring is compromised, the active suspension system will fail completely. Semi-active suspension systems are designed in such a way that if any of the above problems occur, it will revert back to a PaS. In this way, semi-active suspensions have an added safety feature. The most prominent aspect of semi-active suspension systems is the use of a controllable damper, as shown in Figure 2.1(b). Real-time vibration reduction is done through this controllable damper. A commonly used controllable damper is the magneto-rheological damper which utilizes a magneto-rheological fluid.

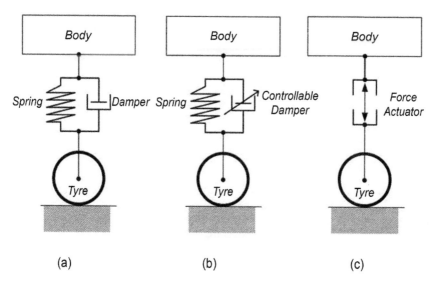

FIGURE 2.1   Suspension systems: (a) passive, (b) semi-active, and (c) active.

2.2   TYPES OF VEHICLE MODELS

In order to analyze the effectiveness of a suspension system, different models have been developed with increasing complexity. The vertical dynamic characteristics of the sprung mass, namely, displacement, velocity, and acceleration, give an understanding of how the sprung mass is moving. This, in turn, determines how comfortable the ride is. The analysis of these dynamic characteristics is done by considering mathematical models of a vehicle. The models are shown as follows:

1. **Quarter Car Model:** This takes into consideration one tire, the unsprung mass connected to the tire, the suspension linkage of that tire, and the sprung mass acting on the tire. In other words, it represents one-fourth of a vehicle. Excitation comes from road disturbances. It is the least complex out of the vehicle models. A schematic diagram of a quarter car model is shown in Figure 2.2.

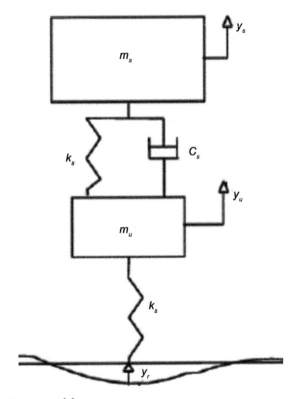

FIGURE 2.2 Quarter car model.

2. **Half Car Model:** These are extensions of quarter car models. A half-car model represents one half of a vehicle split along the longitudinal axis. It is essentially two quarter car models coupled with a rigid element. It takes into account pitching motion along with the vertical motion. A schematic diagram of a half-car model is shown in Figure 2.3.

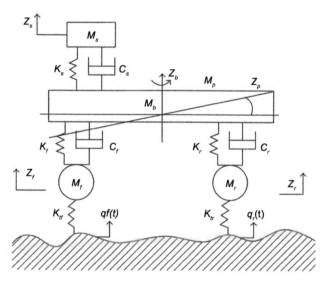

FIGURE 2.3   Half car model.

3. **Full Car Model:** These are extensions of half car models. They represent the dynamic structure of the entire vehicle. It is essentially two-half car models coupled with two-rigid elements at the front and the rear. A schematic diagram of a half-car model is shown in Figure 2.4.

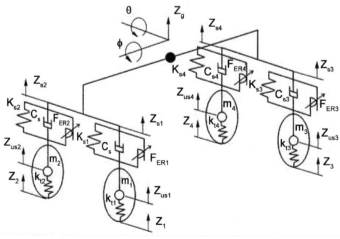

FIGURE 2.4   Full car model.

In this study, the displacement, velocity, and acceleration of the sprung mass of a quarter car model are analyzed to understand the efficacy of the suspension system.

Modeling and simulation of the semi-active quarter car system are done in the Simulink environment of MATLAB.

2.3   MAGNETORHEOLOGICAL (MR) DAMPERS

Magnetorheological (MR) dampers contain MR fluids and are used in semi-active systems. MR dampers have become a hot topic for research due to their intriguing physical features and their ability to control the damping characteristics of a variety of mechanical systems. The construction of the MR damper is similar to a conventional hydraulic damper, with some minor changes. The piston inside the MR damper is an electromagnet. When a magnetic field is applied, the rheological properties of the MR fluid get altered. The yield strength of the fluid increases and this changes the damping characteristics. In this system, damping characteristics are regulated by controlling the strength of the magnetic field. Semi-active systems combine the positive aspects of both active and passive control systems [4].

MR fluids have shown the greatest promise in the application of dampers in the field of semi-active control systems. Small MR dampers are used in seat suspension systems of heavy-duty trucks to prevent unwanted vibrations from reaching the driver. Large MR dampers are used in suspension systems of automobiles. Some MR dampers are specially made to be used in prosthetic devices. In all of these devices, one of the most important fluid properties is low off-state viscosity. MR dampers allow for control of certain properties in real-time, without the use of high-end technology [5].

2.3.1   TYPES OF MR DAMPERS

1. **The Mono Tube Damper:** The monotube MR damper contains a single piston and cylinder. The MR fluid is present in the reservoir and there is an accumulator mechanism to adjust the change in volume that occurs from the movement of the piston rod. The accumulator piston acts as a barrier between the MR fluid reservoir and the compressed gas reservoir. The compressed gas (usually nitrogen) is used to compensate for the change in volume caused

by the movement of the piston rod into the housing. This damper is most commonly used because it is very compact and can be installed in any orientation.

2. **Twin Tube Damper:** This type of MR damper has two fluid reservoirs. As per the Figure, there are two housing: the inner and the outer separated by a foot valve. The inner housing guides the piston rod assembly, in a manner similar to that of the monotube MR damper. The volume enclosed by the inner housing is known as the inner reservoir. The volume present between the inner and outer housing is known as the outer reservoir. MR fluid is present inside the inner reservoir, ensuring that there are no air pockets.

Due to the movement of the piston rod, some change in volume will occur. To compensate for this movement, the outer reservoir is partially filled with MR fluid. Hence, the outer reservoir functions the same way as the pneumatic accumulator in the monotube damper. A valve assembly – known as "the foot valve assembly" – is present at the bottom of the inner housing to control the flow of fluid between the two reservoirs. As the piston rod enters the damper, some MR fluid flows from the inner reservoir to the outer reservoir through the compression valve, which is a component of the foot valve assembly. The volume of liquid displaced by the piston rod as it enters the inner housing is equal to the volume of liquid entering the outer reservoir from the inner reservoir. As the piston rod is retracted from the damper, the MR fluid moves back to the inner reservoir from the outer reservoir through the return valve, which is also a part of the foot valve assembly [6].

2.3.2   APPLICATION OF MR DAMPERS

MR dampers are becoming increasingly sought-after in semi-active vehicle suspension systems because of their mechanical simplicity, high dynamic range, low power requirements, large force capacity, and reliability. Ride quality and ride handling are two very important aspects of vehicle dynamics. They are at constant odds with each other. Good ride quality implies poor ride handling and vice versa. MR dampers offer a compromise solution for these two conflicting properties. The BMW R 1200 GS has an additional MR damper for even smoother running and superior ride ability.

MR dampers offer superior properties that can be incorporated in motorcycle suspension designs. Changing the viscosity of the fluid is the best way to tune a motorcycle suspension. In the case of an MR damper, this is achieved by controlling the applied magnetic field. The range of control is virtually infinite within the off-state and saturation state, making them a perfect substitute for traditional hydraulic shock absorbers for front motorcycle suspension systems [7].

2.4　MATHEMATICAL MODELS FOR MR DAMPER

A number of mathematical models have been developed to describe the nature of MR damper. The models use physical characteristics such as friction to describe the dynamics of the device. The key features of some models are given in the following sub-sections:

1. **Bingham Model:** It is based on the Bingham plastic model. In this model, the body is assumed to behave like a solid until the minimum yield stress is exceeded and then follows a linear relationship between stress and the rate of shear deformation. The Bingham model is shown in Figure 2.5.

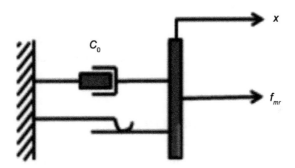

FIGURE 2.5　Bingham model.

The governing equation for this model is:

$$f_{mr} = f_c(\dot{x}) + c_0\dot{x} + f_0 \tag{2.1}$$

where f_{mr} is the force due to the damper, c_0 is the damping coefficient, f_c is the frictional force, $\dot{x}$ is the piston velocity and f_0 is to account for the presence of an accumulator [8].

2. **Hyperbolic Tangent Model:** It has two Voight elements connected by an inertial element that resists motion through the Coulomb friction element. The hyperbolic tangent model is shown in Figure 2.6.

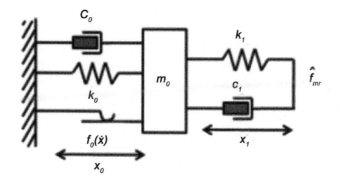

FIGURE 2.6 Hyperbolic tangent model.

The governing equations for this model are:

$$\dot{x}_0 = \epsilon_1 x_0 + \epsilon_2 x + \epsilon_3 f_0 \tanh\left(\frac{\dot{x}_0}{V_r}\right) \tag{2.2}$$

$$f_{mr} = \begin{pmatrix} -k_1 \\ -c_1 \end{pmatrix}^T x_0 = \begin{pmatrix} k_1 \\ c_1 \end{pmatrix}^T \tag{2.3}$$

$$\epsilon_1 = \begin{bmatrix} 0 & 1 \\ -\dfrac{(k_0 + k_1)}{m_0} & -\dfrac{(c_0 + c_1)}{m_0} \end{bmatrix} \tag{2.4}$$

$$\epsilon_2 = \begin{bmatrix} 0 & 0 \\ -\dfrac{k_1}{m_0} & -\dfrac{c_1}{m_0} \end{bmatrix} \qquad\qquad (2.5)$$

$$\epsilon_3 = \begin{bmatrix} 0 \\ -\dfrac{1}{m_0} \end{bmatrix} \qquad\qquad (2.6)$$

where k_1 and c_1 are the spring and dashpot which predict the pre-yield viscoelastic behavior while k_1 and c_1 are the spring and dashpot which predict the post-yield viscoelastic behavior. m_0 is the inertia of the device and f_0 is the yield force. x_0 denotes plastic deformation. V_r is the reference velocity [9].

3. **Dahl Friction Model:** The Dahl model consists of a Coulomb friction element with a lag in the change of friction force when the direction of motion changes. The Dahl model is shown in Figure 2.7.

The governing equations for this model are given as:

$$f_{mr} = k_x(v)\dot{x}(t) + k_w(v)z(t) \qquad\qquad (2.7)$$

$$\dot{z} = p(\dot{x} - |\dot{x}|z) \qquad\qquad (2.8)$$

where k_x and k_w are voltage dependant parameters [10].

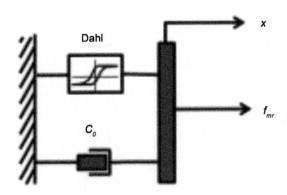

FIGURE 2.7   Dahl friction model.

4. **Bouc-Wen Model:** The Bouc-Wen hysteresis model is one of the simplest and the most effective models to predict the hysteretic behavior of devices. The Bouc-Wen model is shown in Figure 2.8.

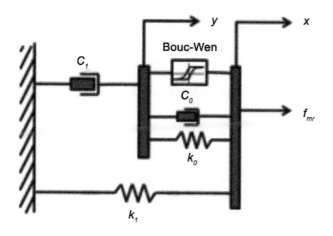

FIGURE 2.8 Bouc-Wen model.

The governing equations for this model are:

$$f_{mr} = k_1(x - x_0) + c_1 \dot{y} \tag{2.9}$$

$$\dot{y}(c_1 + c_0) = k_0(x - y) + c_0 \dot{x} + \alpha z \tag{2.10}$$

$$\dot{z} = -\phi|\dot{x} - \dot{y}|z|z|^{n-1} - \beta(\dot{x} - \dot{y})|z|^n + A(\dot{x} - \dot{y}) \tag{2.11}$$

where $\dot{x}$ is the piston velocity, c_0 is the damping coefficient, k_0 is the stiffness of the device, x_0 is the initial deflection of the spring, x_0 is the initial deflection of the spring and z is the unmeasurable evolutionary variable which accounts for hysteresis behavior. $\alpha, \beta, \varphi, n$ and A are parameters used to control the hysteresis loops. In traditional mathematical models, these parameters are given constant values. To produce an adaptive nature in the device, these parameters should vary based on the external environment. This variation can be produced through fuzzy logic control [11].

The input voltage to the MR damper is governed by the following equations:

$$\alpha = \alpha_a + \alpha_b u \tag{2.12}$$

$$c_0 = c_{0a} + c_{0b} u \tag{2.13}$$

$$c_1 = c_{1a} + c_{1b} u \tag{2.14}$$

where u is the first order filter for the voltage, given by the following equation:

$$\dot{u} = -\eta \left(u - v \right) \tag{2.15}$$

where, v = applied voltage. Here, A change in the voltage will be carried through Eqs. (2.9)–(2.15) to ultimately produce a change in the force of the MR damper. The voltage input to the MR damper is held at a constant value of 5 V. The remaining variables including z are mathematical parameters and do not have any physical meaning. The Bouc-Wen model in Simulink is shown in Figure 2.10.

2.5 SEMI-ACTIVE SUSPENSION SYSTEM MODELING

Simulation of a quarter car suspension model is a very effective method to investigate the influence of damping forces on the response characteristics of a vehicle under different road surface inputs. In this case, the MR damper model (Bouc-Wen model) is incorporated into a quarter car suspension system for subsequent dynamic analysis of the entire system. The quarter car model of Bouc-Wen-based semi-active suspension system (BWSAS) is shown in Figure 2.9.

The quarter car system described here has been modeled with Simulink software for subsequent response analysis of the vehicle due to road disturbance. The quarter car model with a semi-active suspension based on the Bouc-Wen model in Simulink is shown in Figure 2.10. The simulation is carried out to plot vehicle performance for body acceleration, body velocity, and body displacement.

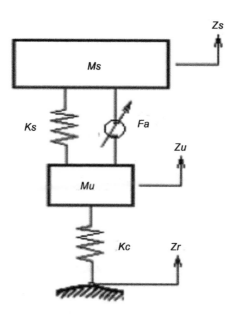

FIGURE 2.9 Bouc-Wen-based semi-active suspension.

2.6 FUZZY LOGIC CONTROLLED SEMI-ACTIVE SUSPENSION

The dynamic model of the fuzzy logic controlled semi-active suspension is given by Figure 2.9 and governed by Eqs. (5.3) and (5.4). The Bouc-Wen model is given in Figure 2.8 and governed by Eqs. (2.12)–(2.15).

To produce effective damping, the voltage input to the MR damper should not be constant. If the voltage is constant, the damping rate would depend on the excitations arising from the road surface. Then this would essentially become a stiffer passive suspension. The voltage should change depending on the road surface. This will give rise to a closed feedback circuit, allowing the force of the MR damper to vary continuously with the road surface. This closed feedback circuit will give the semi-active suspension the ability to adapt to the irregularities of the road surface.

The inputs to the fuzzy logic controller are displacement and velocity of the sprung mass. The output is the applied voltage. The first stage of the controller is the fuzzification of the inputs. Here, the inputs are converted into grades of membership. The Gaussian membership curves are used for the displacement and velocity input variables.

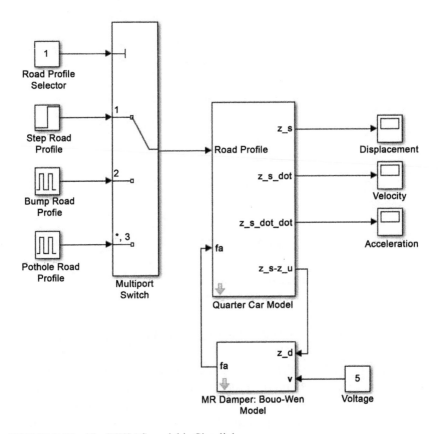

FIGURE 2.10 The BWSAS model in Simulink.

The next stage of the controller is developing the fuzzy inference engine. The Mamdani-type fuzzy inference method is used. Here, a number of rules are made to relate the membership grades of the displacement and velocity to the voltage. These rules are known as fuzzy rules.

The last stage of the fuzzy logic controller is defuzzification. In this stage, the linguistic values of the voltage are converted to crisp values which can be read by the MR damper. This is done by means of a triangular membership function and using the centroid method.

The quarter car model with a semi-active suspension based on fuzzy logic control (FLCSAS) in Simulink is shown in Figure 2.11. There are three scalar gains present, namely G_d, G_{vel}, and G_v. They are for the displacement of the sprung mass, the velocity of the sprung mass, and input voltage to the MR damper. They are assumed as unity.

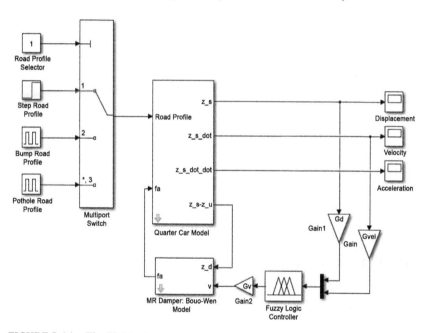

FIGURE 2.11 The FLCSAS model in Simulink.

2.7 RESULTS AND DISCUSSION

The fuzzy logic controlled semi-active suspension is validated for a step road profile and its effectiveness is analyzed by comparing the results of the other models. Optimization of the fuzzy logic controlled semi-active suspension is done and the results are compared with the non-optimized model. Vertical displacement, velocity, and acceleration of the sprung mass of different models will be compared.

2.7.1 ROAD PROFILE

The analysis of all models is done using a step road profile of height 0.1 m with a phase delay of 1 second. The road profile is shown in Figure 2.12.

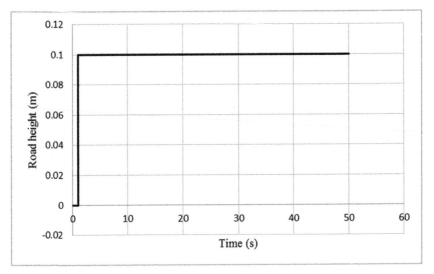

FIGURE 2.12   Step road profile.

2.7.2 RESPONSE TO STEP ROAD PROFILE

In the next phase, passive suspension, BWSAS, and FLCSAS models are compared taking step road profile as the input. The results for vertical displacement, velocity, and acceleration for the passive suspension, BWSAS, and FLCSAS models are shown in Figures 2.13–2.15, respectively. Detail analysis of the PaS is not shown here, only the results are presented.

From Figure 2.13, it is observed that the FLCSAS and BWSAS models have a better response than the passive suspension model. The response of the BWSAS model starts to lag slightly behind the FLCSAS response after 15 seconds. The magnitude of displacement of the FLCSAS and BWSAS are both lower than that of the passive model. The FLCSAS and BWSAS models dissipate the vibrations quicker than the passive system.

Figure 2.14 shows that the FLCSAS and BWSAS model have higher initial velocity values compared to the passive system. The FLCSAS and BWSAS models dissipate vibrations fast and the magnitude of velocities drop quickly.

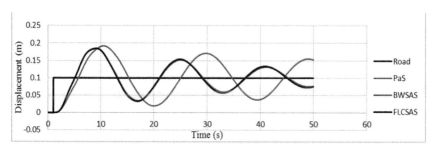

FIGURE 2.13 Displacement response for step road profile.

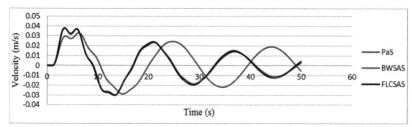

FIGURE 2.14 Velocity response for step road profile.

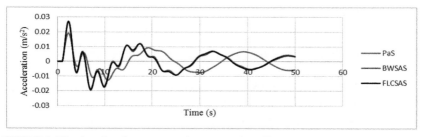

FIGURE 2.15 Acceleration response for step road profile.

However, from Figure 2.15 it is also noticed that the FLCSAS and BWSAS have very high initial values of acceleration. Although the high values start to dissipate quickly, there is too much noise associated with their drop. This noise is very prominent and continues even after 30 seconds.

The analysis of the semi-active suspension systems (BWSAS and FLCSAS) and PaS demonstrates that both the BWSAS and FLCSAS models have superior dynamic characteristics than the PaS. However, BWSAS and FLCSAS show very similar responses. The only difference is that the BWSAS starts to lag slightly behind the FLCSAS after 30 seconds. The BWSAS has reached its saturation state considering 5 V as the input. It cannot be changed any further. The FLCSAS, on the other hand, can undergo further tuning, as explained in the next section.

2.7.3   *OPTIMIZATION FUZZY LOGIC CONTROLLER*

The FLCSAS model analyzed so far did not have any scalar gain. Optimization of the FLCSAS model is done by tuning these values [12]. This is done manually by trial-and-error. A particular value for a scalar gain is initially assumed and the results are analyzed. Another value is assumed for another scalar gain and the results are analyzed. This process is repeated until the best response is achieved. A multitude of values was obtained for these three gains; however, Table 2.1 shows the values which give the best results. The optimized FLCSAS model will be abbreviated as OFLCSAS.

TABLE 2.1   Tuned Values for the Scalar Gain Parameters

Scalar Gains	Tuned Values
G_d	7.2
G_{vel}	6.8
G_v	10

2.7.3.1   *OPTIMIZED RESULTS FOR STEP ROAD PROFILE*

The FLCSAS and OFLCSAS are compared taking step road profiles as the input. The results for vertical displacement, velocity, and acceleration for the FLCSAS and OFLCSAS models are shown in Figures 2.16–2.18, respectively.

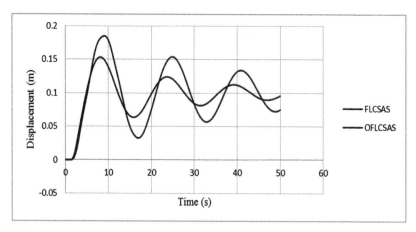

FIGURE 2.16 Optimized displacement for step road profile.

It is observed from Figure 2.16 that the peak values of displacement obtained from the OFLCSAS model are significantly lower than that of the FLCSAS model. The vibrations are dissipated faster as well, providing lesser settling time.

The response of the OFLCSAS model is better than the FLCSAS model (Figure 2.17). The rugged peaks present in the FLCSAS response are absent in the OFLCSAS response, allowing smoother dissipation of the unwanted vibrations.

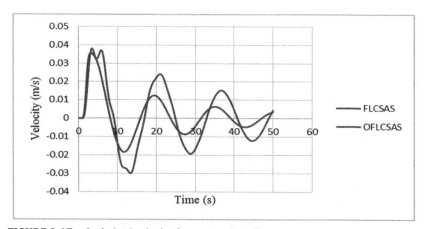

FIGURE 2.17 Optimized velocity for step road profile.

From Figure 2.18, it can be seen that the OFLCSAS model has higher initial acceleration than that of the FLCSAS model. However, this peak value falls very rapidly. The overall acceleration response of the FLCSAS is very jagged, having too many irregularities. The response of the OFLCSAS is far smoother having almost no irregularities. The smoothness of the acceleration profile plays a major role in the comfort of the passengers.

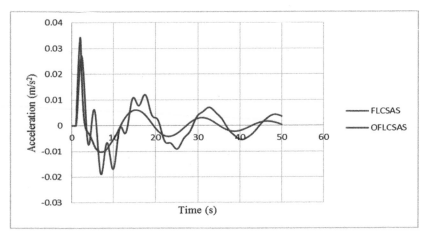

FIGURE 2.18   Optimized acceleration for step road profile.

2.8   CONCLUSION

In this chapter, different types of suspension systems are discussed. The advantage of a semi-active suspension system is highlighted. It is shown that MR damper is an important part of automotive suspension systems and, hence, plays a major role in controlling the dynamic nature of the vehicle. The effectiveness of the MR damper was studied. Initially, a quarter car model with PaS was designed. Another quarter car model was developed with an MR damper as a member of the semi-active suspension system. The behavior of the MR damper was predicted by the Bouc-Wen hysteresis model. The final quarter car model was developed with an MR damper based on the Bouc-Wen model and controlled by a fuzzy logic controller. These three models were analyzed by considering the step road profile. The fuzzy logic controller was optimized by altering the values of the scalar gains to produce better results. The results of the analysis are as follows:

1. The step road profile was considered first. The performance of the models was analyzed by monitoring the displacement, velocity, and acceleration profiles. The FLCSAS and BWSAS models showed an overall better response than the PaS. The FLCSAS and BWSAS models had high initial values of acceleration. The values dissipated quickly, but there was still too much noise present.
2. The OFLCSAS model showed a better response than the FLCSAS model. Since the BWSAS and FLCSAS models showed similar results, by extension, the OFLCSAS model performed better than the BWSAS as well.
3. The only drawback of the OFLCSAS is the slightly higher initial acceleration in all the road profiles. After 5 seconds, the acceleration starts to drop rapidly and the profile generated is much smoother than the profile of the FLCSAS model.

KEYWORDS

- **Bouc-Wen model**
- **engine control unit**
- **fuzzy logic controller**
- **passive suspension system**
- **semi-active suspension**
- **step road profile**

REFERENCES

1. Why is Your Car's Suspension so Important? (2018). *WIYGUL Automotive Clinic.* Conceptual Minds.
2. Wikipedia Contributors, (2018). *Suspension (Vehicle).* Wikipedia, the Free Encyclopedia.
3. Milliken, W. F., & Milliken, D. L., (1995). Race car vehicle dynamics. *Warren Dale: Society of Automotive Engineers.*

4. Gillespie, T. D., (1992). Fundamentals of vehicle dynamics. *Society of Automotive Engineers (SAE)*.

5. Świtoński, E., Mężyk, A., Duda, S., & Kciuk, S., (2007). Prototype magneto rheological fluid damper for active vibration control system. *Journal of Achievements in Materials and Manufacturing Engineering, 21*(1), 55–62.

6. Kciuk, M., & Turczyn, R., (2006). Properties and application of magneto rheological fluids. *Journal of Achievements in Materials and Manufacturing Engineering, 18*(1–2), 127–130.

7. Ashfak, A., Saheed, A., Rasheed, K. K. A., & Jaleel, J. A., (2011). Design, fabrication, and evaluation of MR damper. *International Journal of Aerospace and Mechanical Engineering, 1*, 27–33.

8. Shames, I., & Cozzarelli, F., (1992). *Elastic and Inelastic Stress Analysis Prentice-Hall*. Inc. Englewood Cliffs, NJ.

9. Gavin, H., Hoagg, J., & Dobossy, M., (2001). Optimal design of MR dampers. *Proceedings of the US-Japan Workshop on Smart Structures for Improved Seismic Performance in Urban Regions, 14*, 225–236.

10. Aguirre, N., Ikhouane, F., Rodellar, J., Wagg, D., & Neild, S., (2010). Modeling and identification of a small scale magneto rheological damper. *Proceedings of ALCOSP 2010 Adaptation and Learning in Control and Signal Processing* (pp. 19–24). IFAC.

11. Spencer, Jr. B. F., & Sain, M. K., (1997). Controlling buildings: A new frontier in feedback. *IEEE Control Systems, 17*(6), 19–35.

12. Ross, T. J., (2009). *Fuzzy Logic with Engineering Applications*. John Wiley & Sons.

CHAPTER 3

A Review of the Application of Mechatronics in Manufacturing Processes

T. MUTHURAMALINGAM

Department of Mechatronics Engineering, SRM Institute of Science and Technology, Kattankulathur – 603203, Tamil Nadu, India, E-mail: muthu1060@gmail.com

ABSTRACT

In the present scenario, the importance of enhancing the performances of conventional manufacturing processes is being increased. It is possible only where integrating the mechatronics approaches in manufacturing processes. The adaptation of control algorithms and modification of existing electronics circuits in the manufacturing processes can considerably enhance the performances of the manufacturing technology. The efficacy of the untraditional machining processes can be well enhanced by the mechatronics engineering concepts. In the present study, an endeavor has been made to analyze the various literature related to the adaptation of mechatronics knowledge in manufacturing technology. The responses of the processes with and without mechatronics approaches have been compared and the merits of those systems over the existing systems have been analyzed. The influences of the various control strategies have also been analyzed and compared. It has been observed that the still better research works can be made to increase the efficacy the manufacturing technology.

3.1   INTRODUCTION

All engineering products are being manufactured using various types of manufacturing processes. The various performance measures of the manufacturing technology can always have to be upgraded. Then only it is possible to enhance the quality of the engineering products. The integration of mechatronics with manufacturing technology can lead to developing intelligent manufacturing technology as shown in Figure 3.1.

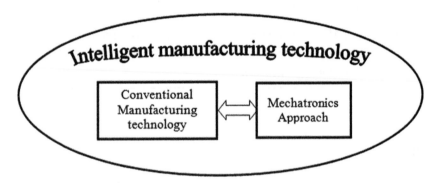

FIGURE 3.1   Mechatronics integration with manufacturing technology.

Mechatronics is a multidisciplinary stream, which can integrate the various domains of technological aspects to enhance the existing technology. The inclusion of sensing technology and controllers along with existing actuators can convert the automatic control of the mechanism of any system. The adaptation of the various intelligent controls and algorithms can further improve the performance measures considerably. The modeling and simulation of the processes and systems can reduce the duration of the design life cycle of the product. It can also further reduce capitalization costs. Software-in-loop (SIL) simulation and hardware-in-loop (HIL) simulation can virtually able to design and develop the product in an efficient way. In the present chapter, the adaptation of the mechatronics approaches with conventional machining, additive manufacturing, and precision engineering has been discussed as shown in Figure 3.2.

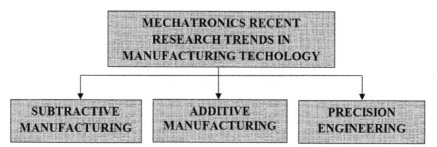

FIGURE 3.2 Recent trends in mechatronics in manufacturing technology.

3.2 MECHATRONICS IN MACHINING PROCESS

The exclusion of the material from the specimens, i.e., subtractive manufacturing can able to produce the engineering product with the required shape and size. The material removal processes can be further classified into the traditional and unconventional machining process. In the traditional machining process, the specimen and the cutting tool are directly making contact with each other. However, the strength of the tool should be higher than the workpiece. In the unconventional machining process, the tool does not touch the specimen during the machining process. Electrical discharge machining (EDM) is one of the important unconventional machining processes in which electrical sparks applied across the machining zone to remove the material by melting and vaporization. The adaptation of mechatronics technology in the machining process can significantly enhance the various performance measures such as material removal rate, surface roughness, surface hardness, tool wear rate, and other considerable machinability parameters in any machining process.

The incorporation of virtual simulation with online real measurements of systems can eliminate fake tool breakdown detection and momentary overloads of the tools during employing the adaptive control. The system has been realized on a CNC machining center for the utilization in production to enhance the efficacy [1]. The improvement in the overall equipment efficacy of machine tools can enhance the resource-efficiency and productivity in any manufacturing process. The adaptation sensors for online monitoring of real systems availability could accomplish the more usefulness and eminence of machining processes. It is necessary to establish an important data source which will facilitate the extraction of error-pattern and the allotment of the pattern to machining processes [2]. The enhancement of energy efficiency involved in machine tools can considerably improve the environmental quality of machining systems. An on-line energy effectiveness monitoring system is essential for the same. Many conservative approaches can supervise energy performance by directly measuring cutting power with torque measuring sensors. A new on-line approach has been developed without the utilization of any torque sensor which can lead to a decreased commissioning cost [3].

The surface roughness of the specimen can be effectively reduced by selecting the proper control algorithms in the EDM process. The iso-energy pulse generator could produce a better surface finish due to the uniform spark energy distribution as shown in Figure 3.3 [4, 5]. The duty factor indicates the time period at which the spark energy applied across the machining zone. The higher duty ratio can lead to higher material exclusion whereas lower duty ratio provides higher surface finish. The automatic switching of the duty factor from higher to a lower value during the roughing to finishing using fuzzy logic controller based mechatronics approach could enhance the machining process in EDM [6]. The tool wear is directly proportional to surface morphology of the machined workpiece in EDM process. The online tool wear monitoring using the mechatronics approach can effectively control the tool wear by implementing the uniform spark energy distribution across the machining zone in machining process as shown in Figure 3.4. The integration of mechatronics approach can reduce the surface roughness in an efficient way [7].

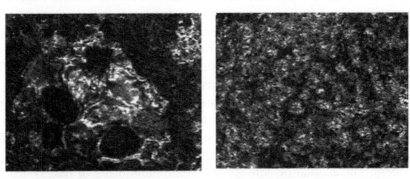

(i) Existing approach   **(ii) Mechatronics approach**

FIGURE 3.3   Effect of mechatronics approach on the machined workpiece surface.

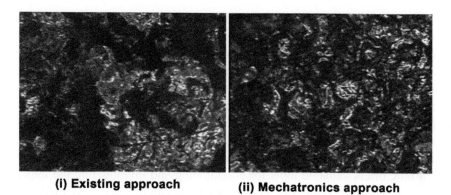

(i) Existing approach   **(ii) Mechatronics approach**

FIGURE 3.4   Effect of mechatronics approach on tool surface after machining.

The melted specimen particles can resolidify on the machined surface as white layer thickness in the EDM process. This layer thickness and distribution can highly contribute to surface hardness in the EDM process. The integration of the intelligent technique along with the existing pulse generator can produce a tiny and uniform recast layer owing to the uniform spark energy distribution in the EDM process. It could reduce the surface hardness of the machined specimen in such a process [8]. The distance between the tool electrode and the specimen in the EDM process has a significant role and nature on the machining characteristics. This air gap can affect the diameter of the plasma column during the machining process. The accumulation of the black carbon layer is developed owing to the arcing effect in the EDM process. The monitoring of the distance using the fuzzy controller with the existing technique could reduce the arcing effect, surface roughness, and waviness in the EDM process [9]. The arcing effect is happened owing to the improper deionization of the dielectric medium in EDM. The stochastic nature of the pulse generator could be considerably reduced by incorporating the mechatronics approach with an existing pulse generator for delivering the spark plasma energy in the EDM process [10].

The pulse shape of the spark pulse generator is one of the recent research trends in the EDM process. The uniform duration of the discharge current pulse could be obtained using the sensor integration of the current sensor and voltage sensor. The higher material removal rate could be obtained using this mechatronics approach [11]. The development of iso-energy pulse generator with the effective control algorithm can enhance the working performance of the pulse generator in the EDM process. The interfacing of the current sensor in series with the specimen-tool combination can considerably control the discharge plasma column in the EDM process [12]. The tool wear is directly proportional to the surface morphology of the machined workpiece in the EDM process. The tool wear could be adjusted by monitoring control flow across the machining zone in the EDM process. The integration of a mechatronics approach can reduce surface roughness in an efficient way [13]. In recent production systems, data-driven system in-process health monitoring is growing in reputation owing to the extensive exploitation of low-cost sensors and their association to the Internet. In the meantime, deep learning offers valuable tools for conditioning and investigating the larger system machinery data [14]. The spark energy can be effectively modified by incorporating the mechatronics approach in the EDM process. The white layer thickness may affect the surface quality of the machined workpiece. The uniform energy can effectively reduce the white layer thickness with uniform in nature [15].

3.3   MECHATRONICS IN ADDITIVE MANUFACTURING

In the machining process, the material is removed as debris, burr, and wastages. Moreover, it can be reused for auxiliary consumption. Nevertheless, the product is formed by the layer by layer formation of specimens under required shapes and size in case additive manufacturing. However, the additive manufacturing processes have to be enhanced in all the possible ways, since the processes have been involved with many input process parameters. The melt pool thermal controller has been designed to control the melt pool high temperature in the constant value, path, and trajectory formulation algorithm is made to preserve a steady glass filament feed direction relative to the scan velocity. The experimentation results exhibit that the melt pool temperature controller can able eliminate the bubble creation over an extensive range of build speeds and that the path planned directional-control is able to obtain better dimensional accuracy throughout closed contours [16]. The micro-scale additive manufacturing μ-AM processes are types of production processes utilized to manufacture micron-sized structures in a series of straight additions of materials as trained by the binary file, as opposed to the lithographic patterning and subtractive etching utilized at existing micro-scale manufacturing. Numerically controlled tools, material addition is an open-loop approach, which necessitates continual user intervention to heuristically tune system parameters [17]. Effective feed-forward control of additive manufacturing of fully dense metallic sections has been proposed. The study is an enhancement of existing control methods of the process, providing high robust and industrially acceptable metrological approaches. The proposed algorithms can successfully moderate these disturbances and sustain the stable deposition while the uncontrolled deposition fails [18]. Two control-oriented models have been proposed to portray the layer-to-layer height evolution of the ink-jet 3D printing technology. While the process has found broad applicability, as with a lot of industrial 3D printing systems, ink-jet 3D printers operated in an open loop, even when sensor measurements are available. One of the primary causes of this is the deficiency of optimal control-oriented models of the process. Two control-oriented models have been developed to eliminate this drawback [19]. Integrated metrology is highly influential for quality control and explanation of micro-scale additive manufacturing and machining processes. The present approach in micro-scale additive manufacturing is to fabricate a component as die. The separate proposed metrology tool has created high production time, lower risking contamination, and lower datum. A new system that integrates electrohydrodynamic jet prints, a promising micro-scale additive production approach, with on-line atomic force microscopy for rapid, die-by-die in-process metrology and performance monitoring [20].

3.4 MECHATRONICS IN PRECISION ENGINEERING

The precision engineering involves various fields of manufacturing techniques such as positional accuracy, precision linear movement, metrology, automatic guided vehicles, and image processing monitoring. The position accuracy and precision movement is a very important parameter in the present scenario of manufacturing technology. The accuracy of the positional can be done using an interconnected cylinder based mechatronics approach. Where one cylinder can be used for coarse movement as that of other cylinders used for fine movement. The precise movement can be evaluated using sensors arrangement. Proportional-integral-derivative (PID) based control in the servo pneumatics can considerably increase the positional accuracy [21]. Even though the speed of response for obtaining linear movement by pneumatics is better than the electrical drives, it is still enhancement needed for obtaining the positional accuracy. The adaptation of the various predictive control and intelligent algorithms can efficiently enhance the positional accuracy with precise movement [22]. The parameters for the interconnected cylinders based servo pneumatic systems should be automatically tuned with optimal process parameters. The better system parameters have been identified using the response surface methodology based on multiple criteria decision making in the servo pneumatics system [23]. The surface roughness is important to surface performance measures. The non-contact-type surface roughness measurement is one of the non-destructive testings which has been used to measure surface performance measures. The principle of a capacitance-based mechatronics approach can be used to forecast the surface performance measures in an efficient way [24]. The sensor integration based mechatronics approach with conventional automated guided vehicles can able to identify the different color pallets in warehouses. These low cost automated guided vehicles can able to perform in well manner [25]. The sensor integration with any conventional approach can enhance the target accuracy in an efficient way. The automatic gun target can be efficiently performed by adaptation of a mechatronics approach [26]. The utilization of line-scan machine vision, 'incremental' image processing based integrated mechatronics design approach can provide better performance than existing frame-based techniques [27].

3.5   CONCLUDING REMARKS AND SUGGESTIONS

The chapter has been framed to analyze the influence of mechatronics approaches in manufacturing processes for enhancing performance measures. The advantages of integrated mechatronic systems over the active systems have been analyzed. It has been inferred that the efficiency of the subtractive, additive manufacturing processes, and precision engineering can be effectively increased by the mechatronics engineering concepts. It has been observed that the still better research works can be made to increase the efficacy the manufacturing technology.

KEYWORDS

- **electrical discharge machining**
- **hardware-in-loop**
- **mechatronics**

- **multidisciplinary stream**
- **proportional-integral-derivative**
- **software-in-loop**

REFERENCES

1. Altintas, Y., & Aslan, D., (2017). Integration of virtual and on-line machining process control and monitoring. *CIRP Annals, 66*(1), 349–352.
2. Emec, S., Krüger, J., & Seliger, G., (2016). Online fault-monitoring in machine tools based on energy consumption analysis and non-invasive data acquisition for improved resource-efficiency. *Procedia CIRP, 40*, 236–243.
3. Hu, S., Liu, F., He, Y., & Hu, T., (2012). An on-line approach for energy efficiency monitoring of machine tools. *Journal of Cleaner Production, 27*, 133–140.
4. Muthuramalingam, T., & Mohan, B., (2014). Performance analysis of ISO current pulse generator on machining characteristics in EDM process. *Archives of Civil and Mechanical Engineering, 14*(3), 383–390.
5. Muthuramalingam, T., & Mohan, B., (2013). Enhancing the surface quality by ISO pulse generator in EDM process, *Advanced Materials Research, 622–623*(1), 380–384.
6. Jaswanth, K., Ashwin, & Muthuramalingam, T., (2017). Influence of duty factor of pulse generator in electrical discharge machining process. *International Journal of Applied Engineering Research, 12*(21), 11397–11399.
7. Muthuramalingam, T., Vasanth, S., & Geethapriyan, T., (2016). Influence of energy distribution and process parameters on tool wear in electrical discharge machining. *International Journal of Control Theory and Applications, 9*(37), 353–359.
8. Muthuramalingam, T., Mohan, B., & Jothilingam, A., (2014). Effect of tool electrode re-solidification on surface hardness in electrical discharge machining. *Materials and Manufacturing Processes, 29*(11–12), 1374–1380.
9. Muthuramalingam, T., Mohan, B., Rajadurai, A., & Saravanakumar, D., (2014). Monitoring and fuzzy control approach for efficient electrical discharge machining process. *Materials and Manufacturing Processes, 29*(3), 281–286.
10. Muthuramalingam, T., Mohan, B., Rajadurai, A., & Antony, M. D., (2013). Experimental investigation of ISO energy pulse generator on performance measures in EDM. *Materials and Manufacturing Processes, 28*(10), 1137–1142.
11. Muthuramalingam, T., & Mohan, B., (2013). Influence of discharge current pulse on machinability in electrical discharge machining. *Materials and Manufacturing Processes, 28*(4), 375–380.
12. Muthuramalingam, T., & Mohan, B., (2013). Design and fabrication of control system based ISO current pulse generator for electrical discharge machining. *International Journal of Mechatronics and Manufacturing Systems, 6*(2), 133–143.

13. Muthuramalingam, T., & Mohan, B., (2013). A study on improving machining characteristics of electrical discharge machining with modified transistor pulse generator. *International Journal of Manufacturing Technology and Management, 27*(1–3), 101–111.

14. Zhao, R., Yan, R., Chen, Z., Mao, K., Wang, P., & Gao, R. X., (2019). Deep learning and its applications to machine health monitoring. *Mechanical Systems and Signal Processing, 115,* 213–237.

15. Muthuramalingam, T., (2019). Measuring the influence of discharge energy on white layer thickness in electrical discharge machining process. *Measurement, 131,* 694–700.

16. Peters, D., Drallmeier, J., Bristow, D. A., Landers, R. G., & Kinzel, E., (2018). Sensing and control in glass additive manufacturing. *Mechatronics, 56,* 188–197.

17. Wang, Z., Pannier, C. P., Barton, K., & Hoelzle, D. J., (2018). Application of robust monotonically convergent spatial iterative learning control to micro-scale additive manufacturing. *Mechatronics, 56,* 157–165.

18. Hagqvist, P., Heralić, A., Christiansson, A. K., & Lennartson, B., (2015). Resistance based iterative learning control of additive manufacturing with wire. *Mechatronics, 31,* 116–123.

19. Guo, Y., Peters, J., Oomen, T., & Mishra, S., (2018). Control-oriented models for ink-jet 3D printing. *Mechatronics, 56,* 211–219.

20. Pannier, C. P., Ojeda, L., Wang, Z., Hoelzle, D., & Barton, K., (2018). An electrohydrodynamic jet printer with integrated metrology, *Mechatronics, 56,* 268–276.

21. Saravanakumar, M. B., & Muthuramalingam, T., (2017). A review on recent research trends in servo pneumatic positioning systems. *Precision Engineering, 49*(1), 481–492.

22. Saravanakumar, D., Mohan, B., Muthuramalingam, T., & Sakthivel, G., (2018). Performance evaluation of interconnected pneumatic cylinders positioning system. *Sensors and Actuators: A. Physical, 274*(1), 155–164.

23. Saravanakumar, D., Mohan, B., & Muthuramalingam, T., (2014). Application of response surface methodology on finding influencing parameters in servo pneumatic system. *Measurement, 54*(1), 40–50.

24. Mathiyazhagan, R., Sampathkumar, S., & Muthuramalingam, T., (2019). Prediction modeling of surface roughness using capacitive sensing technique in machining process. *IEEE Sensors.* doi: 10.1109/JSEN.2019.2927174.

25. Muthuramalingam, T., Mohamed, R. M., Saravanakumar, D., & Jaswanth, K., (2018). Sensor integration-based approach for automatic fork lift trucks. *IEEE Sensors, 18*(2), 736–740.

26. Mohamed, R. M., & Muthuramalingam, T., (2018). Tracking and locking system for shooter with sensory noise cancellation. *IEEE Sensors, 18*(2), 732–735.

27. King, T., (2003). Vision-in-the-loop for control in manufacturing. *Mechatronics, 13*(10), 1123–1147.

CHAPTER 4

Application of a Nature-Inspired Algorithm in Odor Source Localization

KUMAR GAURAV,[1] RAMANPREET SINGH,[2] AJAY KUMAR,[1] and RAM DAYAL[3]

[1]Department of Mechatronics Engineering, Manipal University Jaipur, Dehmi Kalan, Jaipur, Rajasthan, India, E-mails: kumar.gaurav@jaipur.manipal.edu; kumarmuj@gmail.com (K. Gaurav)

[2]Department of Mechanical Engineering, Manipal University Jaipur, Dehmi Kalan, Jaipur, Rajasthan, India

[3]Department of Mechanical Engineering, Malaviya National Institute of Technology, Jaipur, Rajasthan, India

ABSTRACT

A nature-inspired algorithm called hybrid teaching learning particle swarm optimization (HTLPSO) has been applied for odor source localization problem in an indoor environment with mobile agents. An optimization problem is formulated to identify the region of maximum odor concentration. With Gaussian plume model as the simulation environment its effectiveness has been demonstrated. Various set of mobile agents in range of {3-15} are used to find and detect the source of odor with high accuracy and concentration, respectively. Results are presented based on source declaration error, a parameter defining the accuracy in finding the odor source. It has been found that less number of mobile agents are required for fast convergence to detect the odor source.

4.1 INTRODUCTION

Odor source localization (OSL) is an important task, which supports the survival of many species in the animal kingdom, to locate food or their mating partners. Enhanced skills in localizing odor sources can give them an extra edge in a competitive environment where food availability is scarce and the search area is huge. Even for humans it can guarantee applications in various areas to detect materials such as plant matter and drugs in customs; locating unexploded mines and bombs; space exploration; detecting fire in its initial stages; landslides or avalanches; damaged buildings; searching for survivors in earthquakes or quarantine application [1]. One of the major catastrophes happened in 2015, when the fire broke out in a hazardous chemical warehouse at Ruihai International Logistics Co. Ltd., located at Tianjin Port, China. This company deals in the business of warehousing hazardous chemicals. This catastrophic accident resulted in 165 fatalities, 798 injuries, and 8 missing people and caused damage to or destruction of 304 buildings, 12,428 commercial automobiles, and 7533 containers. As much as CNY 6.866 billion direct economic losses were reported due to this accident, the costliest industrial disaster in China in recent years [2]. There-fore, research interest in this field has grown immensely in recent times.

Within active olfaction scope, researchers can develop either of two prominent approaches namely, a single robot and with multi-robots. In comparison with a single robot, the multi-robot strategy may prove to be a better player in terms of adaptability to a complex flow field environment, improved search efficiency, and success rate [3, 4]. Formation based algorithms, let a robot or group of robots maneuver in a particular spatial arrangement to find the target [5]. Multi-robots in OSL were reported long back using a distributed algorithm based on odor and flow information. It was only meant for indoor environment conditions having considerable airflow (>0.05 m/s) [6]. OSL problem has also been addressed in a dynamic environment by modified particle swarm optimization (M-PSO), which can continuously track a changing optimum over time. PSO can be improved or adapted by incorporating the change detection and responding mechanisms for solving dynamic problems. Charge PSO, which is another extension of the PSO has also been applied to solve the dynamic problem [7]. A new M-PSOT algorithm dynamically adjusts two learning factors in the velocity update equation based on the effect of wind on the self-cognition and social cognition of a particle [8]. Modified genetic algorithm (MGA) has been implemented which can calculate preliminary source inversion, with the selection of best candidates through the crossover with eliminated individuals from the population. It is a three-stage solution by identifying the search zone using Markov Chain Monte Carlo (MCMC) sampling, locating the leak source using a modified guaranteed convergence particle swarm optimization (PSO) algorithm [9]. PSO has been extensively used in OSL until it found that robots have a tendency to move towards the same local optimum (no odor source presence) [10], leading to a slow convergence rate and finally fail to declare the source due to reduced mobility. Lack of collaboration is another factor; mobile agents do not directly take concentration measurements of other agents into account [11].

The drawbacks in PSO have been eradicated to some extent by various strategies proposed such as adjusting the learning factors and the inertia weight. [12–14]. PSO has also been modified in which two learning factors kept updating based on the wind [8].

Apart from modifications in parameters of PSO, there are other efforts also to combine different methods with PSO to achieve hybridization. An approach for OSL utilizes the strength of both PSO and bacterial foraging optimization (BFO). Two new operations namely elimination dispersal and chemotaxis are combined with PSO. To avoid stagnation in local optima elimination dispersal is used whereas chemotaxis helps in tracing the plume area. The simulation was done to validate the approach [15]. The gravitational search algorithm and PSO may be hybridized for OSL simultaneously using the concept of the search counter [16]. Another approach could be, concatenating grey wolf optimizer with PSO [17]. The hybridization of PSO has not seen much work in recent times. In contrast to these algorithms in which parameter tuning is required, a parameterless algorithm known as teaching-learning based optimization may be used [18]. Nevertheless, its hybridized version may also be used. It's reported that hybrid teaching-learning PSO outperforms the well-established benchmark algorithms synthesis of mechanisms generate path [19]. As it is a novel fused algorithm, we have implemented for OSL and its effectiveness is demonstrated through simulation experiments.

4.2 ENVIRONMENT FOR SIMULATION

A controlled environment based on the advection-diffusion model [20, 21] is set up to test the developed algorithm. The mathematical expression for concentration 'C' of chemical gas at any point x = (x, y) in the domain of interest is given by:

$$C(x,\infty,x',y') = \frac{q_o \exp\left(-\dfrac{v(d-x+x')}{2k}\right)}{2\pi kd} \tag{4.1}$$

where 'q_o' is chemical release rate, 'v' is the wind velocity in the positive direction of x-coordinate, is Euclidean distance between the point source and the point of interest, 'k' is the diffusion coefficient and (x', y') are the coordinates of the point source. The layout of the arena is 20 m × 20 m enclosed square area considered very similar to a large room in a building. Virtual agents (VA) start its search operation from random points at the beginning of the simulation run. Other parameters, which produced as shown in Figure 4.1 are furnished in Table 4.1.

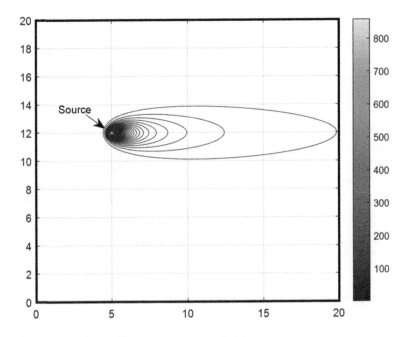

FIGURE 4.1   Simulation environment with the source at (5,12).

TABLE 4.1 Parameters That Produced Figure 4.1

Parameters →	*v*	*k*	*q*	*x'*	*y'*
Values	0.03 m/s	0.005	0.5 mg/s	5 m	12 m

4.3 ALGORITHM IMPLEMENTATION

In this section, the implementation of the HTLPSO algorithm is discussed. The HTLPSO is an ensemble of two well-established nature-inspired optimization algorithms, namely, TLBO, and PSO. In HTLPSO, the teaching phase of TLBO is combined with the PSO while maintaining the size of the initial population and the best solutions are kept as it is. The flowchart of HTLPSO is depicted in Figure 4.2.

The algorithm combines the random population (even) generated at the beginning. The generated population is utilized by both TLBO and PSO. Thereafter, the best half of the population obtained after PSO is combined with the best half of the population obtained after the teaching phase of the TLBO to assure that size of the population remains the same. The population so obtained is utilized for the learning phase of TLBO. The completion of the first iteration is indicated by the end of the learning phase. Thereafter, the final population obtained after the learning phase is utilized for the next iteration. The ensemble of TLBO with PSO helps in improving the computation speed and efficiency of the algorithm.

4.4 EXPERIMENTAL SETUP AND VALIDATION

4.4.1 OUTLINE OF SIMULATION EXPERIMENTS

Simulation experiments are designed for two categories. Each category has been made based on the number of iterations. In the first category all, the experiments are conducted with a maximum of 60 iterations whereas the second category consists of experiments with 120 iterations. In each category, there are 13 sets. Each set corresponds to a team size of VAs. For example, the first set corresponds to the VAs team size of three likewise VAs team size of 15 corresponds to set 13. In total, there are 26 sets. For a particular set, there are 100 simulation experiments to incorporate stochastic disturbances. Details of simulation time for both categories have been furnished in Table 4.2. This simulation time is not comparable to real-time. To evaluate the performance of the adopted algorithm, two factors are taken into consideration. First is the success rate and the second one is average no. of iterations consumed. The next section describes source declaration error (SDE) based on which success rate has been defined.

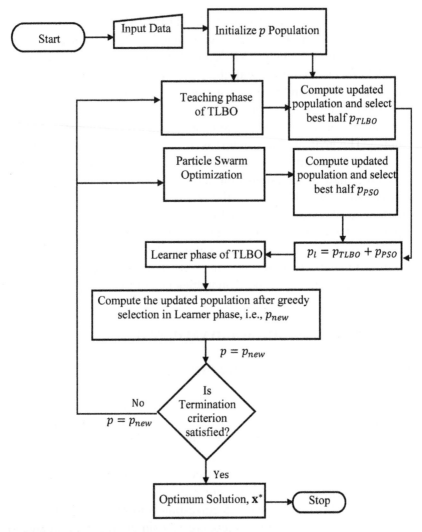

FIGURE 4.2   Flowchart of HTLPSO algorithm.

TABLE 4.2 Details of the Simulation Experiments with VAs Team Size of 3 to 15 with Run Time

Sl. No.	No. of Agents	No. of Iterations/ Experiment	Total Time (secs.): 60 Iterations	Total Time (secs.): 120 Iterations	No. of Simulation Experiments
1.	3	60	1.97	2.40	100
2.	4	60	2.085	2.56	100
3.	5	60	2.26	2.67	100
4.	6	60	2.53	3.11	100
5.	7	60	2.84	3.02	100
6.	8	60	2.59	3.23	100
7.	9	60	2.66	3.23	100
8.	10	60	2.73	3.38	100
9.	11	60	2.86	3.71	100
10.	12	60	3.06	3.83	100
11.	13	60	3.18	4.05	100
12.	14	60	3.26	4.20	100
13.	15	60	3.37	4.40	100

4.4.2 SOURCE DECLARATION ERROR (SDE)

"Source declaration error" (SDE) is a term introduced in this section to have a better understanding of the success rate. In an experiment when the agents close into the source but declare/identify it nearby other than the actual location, at that time Euclidean distance from the global best agent to the target location is computed. It is termed as SDE. In the present case, the source position is fixed at (5,12), therefore relative to his position SDE is calculated and it is a very important parameter which decides accuracy in declaring the source successfully.

4.4.3 SUCCESS RATE

In this subsection, the successful run has been defined as the closest reach of the global best agent to the odor source. Therefore, the VA who reaches less than 0.02 meters in the vicinity of the source can be said to have successfully declared the source and named SDET1. Experiments are also carried out for SDE at 0.05 meters named as SDET2. Hence the success rate is:

$$S = \frac{R}{T} \tag{4.2}$$

where: R = successful runs; and T = total runs.

4.5   RESULTS AND DISCUSSIONS

Simulation time for both categories namely 60 and 120 iterations have been tabulated in Table 4.2. It can be observed that for 100 simulation experiments, simulation time is only 1.97 secs for 60 iterations for VAs team size of 3 whereas it is 2.40 secs for 120 iterations. Likewise, all the run time pertaining to a different team size of VA can be referred from Table 4.2. It shows that the convergence of HTLPSO is very fast. A better picture will come from average no. iterations required. Figure 4.3 shows the standard deviation plot of SDE vs. no. of VA when maximum no. of iterations is 60 only. Minimum SDE in an experiment can be extracted out of 60 values. Therefore, for 100 runs a standard deviation of 100 SDEs is calculated which includes both successful and unsuccessful experiments. In Figure 4.3 it can be observed that SDE is higher, i.e., 3.06 when no. of agents are only 3 but monotonically decreases with no. of agents increasing up to 5. After that, for no. of agents 6 to 15, it immediately becomes zero by showing very high accuracy in source declaration/identification. It is noticeable that for successful experiments, the standard deviation for SDET1 is 0.0048, and SDET2 is 0.0114 for VAs team size of 3 shown in Figure 4.4. For both SDET1 and SDET2, the standard deviation is much lower than SDE (unsuccessful and successful experiments) shown in Figure 4.3 for 100 experiments. It can be concluded from both Figures 4.3 and 4.4 that standard deviation converges to zero for VAs team size of 6 or higher.

   In Figure 4.5, the success rate has been shown for 100 simulation experiment times for different no. of VA. For VAs, a team size of 3 only 25 times out of 100 sources has been identified/declared. It has increased to 68 for VAs team size of 4 and further to 98 when it increased to 5. For VAs team size of 6 to 15, in each case, a 100% success rate has been achieved.

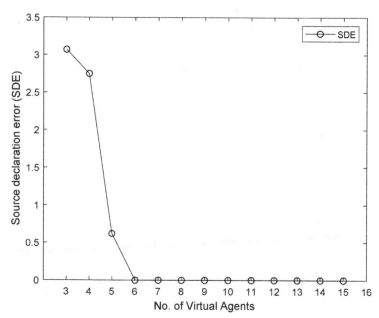

FIGURE 4.3   Standard deviation plot of SDE for virtual agents.

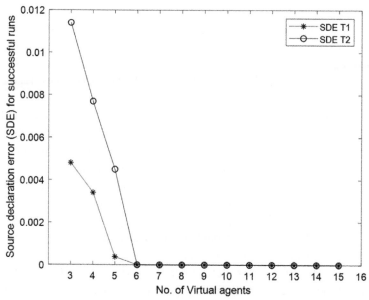

FIGURE 4.4   Standard deviation plot of SDE with virtual agents only for successful experiments.

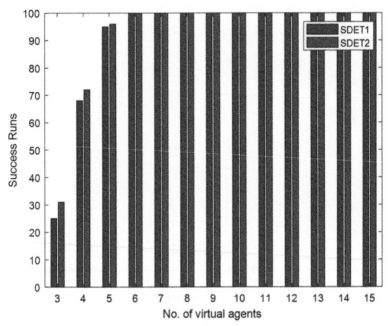

FIGURE 4.5   Success rate with no. of virtual agents.

It is an attempt to report average iterations utilized for this process in successful runs only. In Figure 4.6 it is observed that for VAs team size of 3, on an average 24.64 iterations are used with a standard deviation of 14.02 in 25 successful runs. Similarly for VAs team size of 4 and 5, on an average 21.7 and 16.11 iterations are utilized respectively. It is to be noted that the success rate for VAs team size of 4 and 5 is 68 and 95 respectively. The standard deviation of iterations in successful experiments, i.e., 25 is 14.02 for VAs team size of 3; it has been shown in Figure 4.6. The standard deviation of iterations in successful runs decreases as the no. of VA increases. But both, i.e., average iterations and standard deviation becomes almost constant when no. of VA is greater than 12. Hence, the maximum no. of VA has been limited to 15 in OSL observing no better results further. Even as we are getting a 25% success rate with on average 24.64 iterations there are chances to increase SDE to reach a 100% success rate and compromising on SDE.

Figure 4.7 presents the box plots of iterations with no. of VA in successful experiments. It gives an overview in one plot, i.e., the maximum and minimum value of iteration, outliers as well as the mean and median. It is evident that for no. of VA greater than 12, iterations used are almost constant.

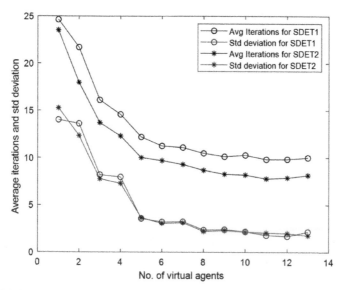

FIGURE 4.6 Average iterations and std. deviation of iterations in 100 experiments with no. of virtual agents.

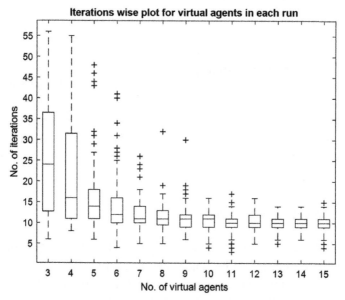

FIGURE 4.7 Iterations details in successful experiments for SDET1 with no. of virtual agents.

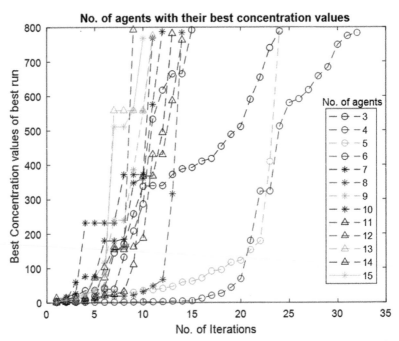

FIGURE 4.8   Maximum concentration values of the best result for SDET1 out of 100 experiments with VAs team size of {3–15}.

The maximum concentration values of the best result out of 100 experiments are plotted with respect to a number of iterations. Figure 4.8 shows that when no. of agents is less than 6, i.e., for 3, 4, and 5 more no. of iterations are required to reach the maximum value whereas for higher no. of VA less no. of iterations are required. It can be clearly observed in Figure 4.8 that, clusters are there separating the no. of VA in two groups, i.e., {3–5} and {6–15}. Indeed, with a larger no. of VA 100% success rate can be achieved with less no. of iterations.

4.5.1   CONVERGENCE PLOTS

Figure 4.9 shows the sequence of convergence while three VA are in pursuit of the odor source. It can be noted that at the first iteration to the third iteration their location changes in line with the wind direction as a result of fast communication with each other. Similarly, from 3rd iteration to 5th iteration they start following the plume centerline moving in collaboration towards the source. At 10th iteration itself, they reach in the proximity to the source but take some iterations to declare/identify the source successfully. Figure 4.9(e) shows the final positions of three mobile agents finally converged at (5,12) means the source was successfully declared/identified. Similarly, in Figure 4.10 it is observed that with the VAs team size of 7, the pursuit lasted for few seconds only and the source is identified/declared immediately after the beginning of the simulation experiment.

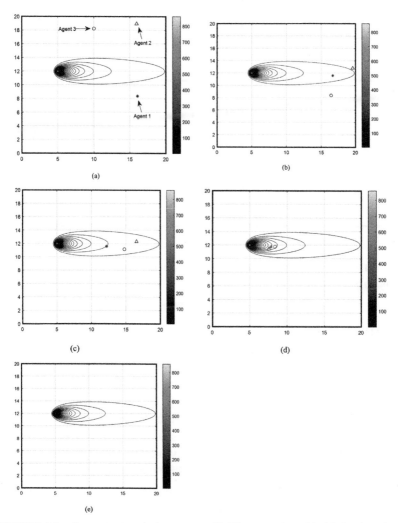

FIGURE 4.9 Convergence to odor source with VAs team size of 3: (a) first iteration, (b) 3rd iteration, (c) 5th iteration, (d) 10th iteration, and (e) 20th iteration.

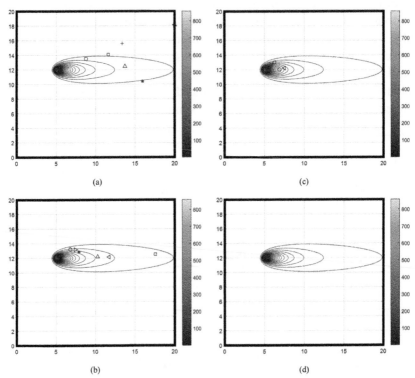

FIGURE 4.10 Convergence to odor source with 7 virtual agents: (a) first iteration, (b) 3rd iteration, (c) 5th iteration, (d) 10th iteration, and (e) 20th iteration.

4.6 CONCLUSION

Hybrid teaching-learning PSO for OSL in an indoor environment has been implemented with the Gaussian plume model as a simulated environment. The highlight of HTLPSO is very fast convergence with high accuracy in declaring the source of emission. Keeping the threshold as low as 0.02 meters, a 25% success rate was achieved with VAs team size of 3. Increasing the threshold can definitely increase the success rate. An important fact is that very less iteration were used to converge. For VAs team size of {3–5}, more iteration is required as compared to team size {6–15} to reach the maximum value. The standard deviation of iterations kept on decreasing with increasing no. of VA. When no. of iterations is increased, SDE decreased leading to higher accuracy. However, it didn't affect the success rate. Implementation in other plume models can be another application of this algorithm.

KEYWORDS

- **Gaussian Plume model**
- **hybrid teaching-learning particle swarm optimization**
- **indoor environment**
- **multi-robot**
- **odor source localization**
- **particle swarm optimization**

REFERENCES

1. Kowadlo, G., & Russell, R. A., (2008). Robot odor localization: A taxonomy and survey. *The International Journal of Robotics Research, 27*, 869–894.
2. Zhou, L., Fu, G., & Xue, Y., (2018). Human and organizational factors in Chinese hazardous chemical accidents: A case study of the '8.12(Tianjin Port fire and explosion using the HFACS-HC. *International Journal of Occupational Safety and Ergonomics, 24*, 329–340.
3. Marjovi, A., & Marques, L., (2013). Optimal spatial formation of swarm robotic gas sensors in odor plume finding. *Autonomous Robots, 35*, 93–109.
4. Chen, Y., Cai, H., Chen, Z., & Feng, Q., (2017). Using multi-robot active olfaction method to locate time-varying contaminant source in indoor environment. *Building and Environment, 118*, 101–112.
5. Soares, J. M., Aguiar, A. P., Pascoal, A. M., & Martinoli, A., (2016). A graph-based formation algorithm for odor plume tracing. In: *Distributed Autonomous Robotic Systems* (pp. 255–269). Springer.
6. Hayes, A. T., Martinoli, A., & Goodman, R. M., (2002). Distributed odor source localization. *IEEE Sensors Journal, 2*, 260–271.
7. Jatmiko, W., Ikemoto, Y., Matsuno, T., Fukuda, T., & Sekiyama, K., (2005). Distributed odor source localization in dynamic environment. In: *IEEE Sensors* (p. 4).
8. Gong, D. W., Qi, C. L., Zhang, Y., & Li, M., (2011). Modified particle swarm optimization for odor source localization of multi-robot. In: *2011 IEEE Congress of Evolutionary Computation (CEC)*, pp. 130–136.
9. Wang, J., Zhang, R., Yan, Y., Dong, X., & Li, J. M., (2017). Locating hazardous gas leaks in the atmosphere via modified genetic, MCMC and particle swarm optimization algorithms. *Atmospheric Environment, 157*, 27–37.

10. Yang, X., Yuan, J., Yuan, J., & Mao, H., (2007). A modified particle swarm optimizer with dynamic adaptation. *Applied Mathematics and Computation, 189*, 1205–1213.

11. Chen, X. X., & Huang, J., (2018). Odor source localization algorithms on mobile robots: A review and future outlook. *Robotics and Autonomous Systems*.

12. Zheng, Y. L., Ma, L. H., Zhang, L. Y., & Qian, J. X., (2003). On the convergence analysis and parameter selection in particle swarm optimization. In: *Proceedings of the 2003 International Conference on Machine Learning and Cybernetics (IEEE Cat. No. 03EX693)* (pp. 1802–1807).

13. Eberhart, R. C., & Shi, Y., (2001). Tracking and optimizing dynamic systems with particle swarms. In: *Proceedings of the 2001 Congress on Evolutionary Computation (IEEE Cat. No. 01TH8546)* (pp. 94–100).

14. Eberhart, R. C., & Shi, Y., (2000). Comparing inertia weights and constriction factors in particle swarm optimization. In: *Proceedings of the 2000 Congress on Evolutionary Computation. CEC00 (Cat. No. 00TH8512)* (pp. 84–88).

15. Zhang, Y., Zhang, J., Hao, G., & Zhang, W., (2015). Localizing odor source with multi-robot based on hybrid particle swarm optimization. In: *Natural Computation (ICNC), 2015 11th International Conference* (pp. 902–906).

16. Jain, U., Godfrey, W. W., & Tiwari, R., (2017). A hybridization of gravitational search algorithm and particle swarm optimization for odor source localization. *International Journal of Robotics Applications and Technologies (IJRAT), 5*, 20–33.

17. Jain, U., Tiwari, R., & Godfrey, W. W., (2018). Odor source localization by concatenating particle swarm optimization and grey wolf optimizer. In: *Advanced Computational and Communication Paradigms* (pp. 145–153). Springer.

18. Rao, R. V., Savsani, V. J., & Vakharia, D., (2011). Teaching–learning-based optimization: A novel method for constrained mechanical design optimization problems. *Computer-Aided Design, 43*, 303–315.

19. Singh, R., Chaudhary, H., & Singh, A. K., (2017). A new hybrid teaching-learning particle swarm optimization algorithm for synthesis of linkages to generate path. *Sādhanā, 42*, 1851–1870.

20. Matthes, J., Groll, L., & Keller, H. B., (2005). Source localization by spatially distributed electronic noses for advection and diffusion. *IEEE Transactions on Signal Processing, 53*, 1711–1719.

21. Cao, M. L., Meng, Q. H., Wu, Y. X., Zeng, M., & Li, W., (2013). Consensus based distributed concentration-weighted summation algorithm for gas-leakage source localization using a wireless sensor network. In: *Control Conference (CCC), 2013 32nd Chinese* (pp. 7398–7403).

CHAPTER 5

Tire Pressure Monitoring Systems: A Case Study in Automotive Mechatronics

NARAYAN KUMAR and B. RAVINDRA

Department of Mechanical Engineering, IIT Jodhpur, Karwar, Jodhpur, Rajasthan, India, E-mail: ravib@iitj.ac.in (B. Ravindra)

ABSTRACT

Sensors are an integral part of mechatronic systems. It is common to introduce sensor characteristics such as range, span, error, accuracy, sensitivity, hysteresis, non-linearity, stability, dead band, resolution, etc., in a basic course on mechatronics. The static and dynamic responses of sensors are also explained in this context. Introducing real-life case studies, while teaching sensors, can motivate students. Recent advances in sensors include wireless transmission and energy harvesting from ambient sources. An interesting case study to understand these developments is the tire pressure monitoring sensor (TPMS). Underinflated tires can lead to accidents. Reduction in tire pressure may also increase wear out of tires and reduction in fuel economy. Tire pressure monitoring is a solution to address these issues. TPMS is made mandatory in some countries. In the existing implementation of TPMS in automobiles, the energy source is a CR battery. Battery operated TPMS have certain limitations. Frequent replacement of the battery is a concern for the user. The idea of using vibrational energy of the tire to power TPMS has been proposed in the literature as an alternative to batteries.

5.1   INTRODUCTION

Sense and respond features are ubiquitous in mechatronic systems. The selection of physical quantity to be sensed is very important in their design. The availability of commercially off the shelf (COTS) sensors made this task simple in recent years. Sensors are often classified based on the input stimuli such as temperature sensor, pressure sensor, humidity sensor, etc. Another classification is based on an analog or digital sensor. The nature of the transduction of the input stimuli to a convenient electrical signal is a critical step. The noisy signal from the sensor is passed on to the signal conditioning circuit. The output of this circuit is fed to the data acquisition (DAQ) system. The following classification of sensors is often used in literature:

1. The physical principle on which the sensor works such as Hall Effect, piezoelectric effect, Seebeck effect, etc.
2. Whether the sensor is a contact/non-contact type.
3. Whether the sensor is wired or wireless.
4. If the sensor requires energy to operate then it is referred to as active else passive.

Sensor selection involves in a clear understanding of the technical specifications given by the manufacturer. The following attributes need to be considered in this regard:

1. Sensitivity of the sensor which is the smallest detectable change in the output of the sensor to a change in the input, evaluated at a specific input.
2. Gain of the sensor defined as the absolute value of the output signal relative to the absolute value of the input signal.
3. Full scale output that specifies the working range of the measured physical quantity.
4. Span of the sensor which is the difference between electrical sensor output values corresponding to the max and min values of the electrical signal.
5. Threshold of the sensor is the minimum value of the input signal needed to stimulate or activate the sensor output.
6. Resolution refers to the incremental change in the input signal that would cause a corresponding change in the sensor output.

7. Response time of the sensor.
8. Linearity of the sensor.
9. Repeatability and Accuracy of the sensor.
10. Precision, Error, and uncertainty of the sensor.
11. Dynamic response of the sensor.

It is typical in mechatronics courses to introduce various kinds of sensors to the students. A typical reference material for this task is the IEEE Handbook of sensors. An alternate pedagogical approach is to present a case study on sensors and emphasize the system integration and design aspects from a mechatronic point of view. It may even be important to expose students to regulatory aspects of engineering systems and how mechatronics and sensors can play a role in meeting the requirements of society. The case of tire pressure monitoring sensors (TPMS) is discussed in this article from this perspective.

5.2   NEED OF TPMS

Most of the drivers are not experienced enough to judge whether tire pressure is correct or not according to specifications. On average, there are ten centimeters thicker air and tire layers between rim and ground. Because of the underinflated tires, there has been a significant number of accidents occurring each year. One of the common causes of accidents is the failure of the tire due to lower pressure. Even if lower pressure may not cause tire failure it can still cause a fatality. In under-inflated tires, the driver does not get good control and may also lead to an increase in stopping distance. The NHTSA (National Highway Traffic Safety Administration) report estimates that the tire pressure sensor system in vehicles would prevent several fatalities.

The contact patch of tire increases due to under-inflated tires, so the effective radius of tire reduces. This way mileage and fuel economy of the vehicle may also get affected. For example, tires that have 6psi lower than the recommended pressure can cause a 5% decrease in fuel economy. That is if any vehicle has a mileage of 20 km per liter then due to under-inflation of 6 psi causes 19 km per liter. In the long term, it is a significant loss to the economy. Not only mileage and fuel economy are affected due to under-inflation of tires but also tires wear very fast closer to sides. The tire expert Rastetter estimated that the tire wears out 25%

faster if the pressure is 6 psi below the specification [1]. It is still a challenge to save this loss in terms of tire wear and fuel economy. This loss is primarily due to the lack of monitoring of the tire pressure. TPMS (tire pressure maintenance system) is the best technological option that alerts the possibility of underinflated tires to the driver. Tires typically lose about one psi of pressure each month. This depends on environmental conditions as well. The seasonal changes and driving patterns also may play a role.

It is interesting to take a look at the regulatory aspect of TPMS worldwide. As per a report of freescale, in the United States FMVSS138 mandates TPMS for new vehicles starting from 2005. In the European Union EC661–2009 mandates TPMS for all new vehicles starting from November 2014. In South Korea and Japan TPMS is mandated on passenger cars from 2013 for new models. In Russia, Kazakhstan, Belarus, Indonesia, Israel, Malaysia, Philippines, TPMS may be required for all new vehicles starting November 2014. Similar efforts are underway in China as well. Thus, there is a strong impetus from the regulatory authorities to install TPMS in all new vehicles.

5.3   TYPES OF TPMS

At present, TPMS can be divided mainly into two types: one of them is based on the wheel speed (also referred to as indirect TPMS). This system compares the differences of the speed values between the tires by the ABS wheel speed sensor system of the vehicle to achieve the purpose of monitoring the tire pressure. The main drawback of this system is that it can't judge the situation when more than two tires are under normal pressure. If the vehicle speed is over 100 km/hr, a large error may result. Thus, only when the vehicle speed is below a certain threshold value, indirect TPMS seems to be useful.

The second option is based on the pressure sensor (also referred to as direct TPMS). This TPMS system makes use of a pressure sensor which is installed in each tire of the vehicle to measure the tire pressure and that data is shown in driver's display and monitors the pressure of each tire. These days many cars are fitted with tire pressure sensors. Rapid miniaturization of sensors is an important development in the deployment of TPMS. The four sensors are connected in a wireless sensor network and the tire pressure information is displayed wirelessly. The main goal is to reduce the power consumption of the wireless sensors. This chapter deals with only direct TPMS as described in the next section.

5.3.1 DIRECT TPMS

A direct tire pressure monitoring system refers to a pressure sensor directly mounted on the wheels or tires of a vehicle. The pressure information is often transmitted to the vehicle using radio frequency (RF) technology. Typical specification of a Microchip TPMS is given below:

- Operating voltage: 2.3–3.3 V.
- Low voltage alert threshold: 2.3 V.
- Stand by current: ~12 µA.
- UHF transmitting frequency: 433.92 MHz.
- UHF transmission baud rate (TE): 100, 200, 400, 800 µs selectable (system default is 400 µs).
- UHF range: ~10 meters.
- LF frequency: 125 kHz.
- LF input sensitivity: ~3 mVPP.
- LF range: Up to 3 meters.
- Pressure sensor type: Analog.
- Pressure sensor range: 1–7 bars absolute.
- Pressure sensor temperature range: 40–125°C.

A direct TPMS is often equipped with the following components in the vehicle:

1. A direct TPM sensor built-in to the back of the valve stem on each one wheel.
2. A TPM warning light.
3. Unique identifier for the tires providing the direction of rotation and speed.
4. A microcontroller unit.
5. Antenna.
6. Batteries.

In existing TPMS a 3 V NiHM (CR2320) battery is often used. Battery operated TPMS have certain limitations and owner's dissatisfaction with battery replacement is a concern. Piezoelectric based tire vibration harvesting is an option in this case.

5.4   CHALLENGES IN TPMS

TPMS must be designed to meet the following requirements:

- The pressure sensor must be small enough because it should be fit inside the tire.
- A small power source to supply power to the sensor.
- The amount of data from the sensor (and the range) that should be sent to the driver's display.
- The environmental influences on sensor data and reliability of sensors.

5.5   ENERGY HARVESTING THROUGH PIEZO BIMORPHS

The idea of using piezoelectric harvesting of tire vibrations to power TPMS is discussed here. A discussion of other challenges outlined above can be found in the relevant literature. In the existing technology, the battery is used to power the sensor. But batteries have limitations in terms of total discharge period and replacement issue. Can batteries as power sources be replaced with piezoelectric energy harvester to meet the requirement of power for the sensor? This idea is around for over a decade. Several prototypes have been made but to the best of the author's knowledge, no commercial solution seems to have been deployed in real vehicles. The space constraints and the tuning of the piezo harvester to the tire vibrations seem to pose some hurdles.

Energy harvesting through piezoelectric materials in the form a bimorph that can extract power from ambient vibrations has been thoroughly researched in the literature. These materials exhibit electromechanical coupling between mechanical and electrical domains. Energy harvested is high when the piezo bimorph is excited at the resonance frequency. Various factors affecting the natural frequency of the piezoelectric bimorph are studied in the literature. Various piezoelectric materials, substrates, and geometrical conditions are compared for energy harvesting from vibrations to power TPMS in the literature. The commercial feasibility of this option needs to be compared with existing battery-operated tire pressure monitoring systems.

5.6 PIEZOELECTRIC ENERGY HARVESTER

A schematic diagram of the bimorph structure for the TPS module is shown in Figure 5.1. A piezoelectric energy harvester is a cantilever beam of piezoelectric plates and metallic plates bonded tightly. As shown in Figure 5.1, the bimorph is fixed at one end and free at another end. The free end vibrates transverse to the length direction, to produce power.

The harvester fits in the shell, so that the bender will not get affected from its surroundings and in the shell, the bender can vibrate freely. The following are the basic design elements for the module:

- Selection of piezoelectric material;
- Selection of metallic layer;
- Harvester is fixed at one end. An insulating mount is required to fix the bender from one end;
- Harvester is free from another end, so that it can vibrate freely. Sufficient free space should be in the shell for the vibration of free end freely;
- The harvester is embedded in a shell. So, a shell or package should be designed to protect the bender as shown in Figure 5.2.

This piezoelectric energy harvester comprises of two piezoelectric layers and three metallic layers. The proof mass is provided to tune the natural frequency of harvester. The whole assembly is embedded in a cylindrical shell.

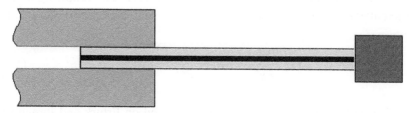

FIGURE 5.1 Piezo bimorph with a tip mass.

FIGURE 5.2   Energy harvester embedded in the shell.

5.7   DESIGN OF TPMS WITH PZT BIMORPH HARVESTER

The challenges are (i) to know the natural frequency of the tire; actually, it can vary very randomly due to change of load, change of speed, change of roads, etc., (ii) design the bimorph which has natural frequency near to the natural frequency of the tire, (iii) design of electrical circuit which will supply undisrupted power to the sensor and required voltage to the sensor, (iv) compact and solid casing for the whole TPMS.

The natural frequencies can be estimated from the solution of the governing equations of the vibrating cantilever beam with a tip mass. The tip mass selection can be made to bring the natural frequency down to match the tire lateral vibrations. The beam with no tip mass needs a distributed parameter mathematical model. The beam with large tip mass can approximate a lumped mass model. Therefore, It is necessary to overview the governing equations of the distributed parameter base excitation model. The standard Euler-Bernoulli beam assumptions (neglecting shear deformation and rotary inertia) for a uniform cantilever beam are used here. The following partial differential equation describes the free vibrations of the beam (Figure 5.3).

$$YI\frac{\partial u(x,t)}{\partial x^4} + m\frac{\partial^2 u(x,t)}{\partial t^2} = 0 \qquad (5.1)$$

The mathematical model used here is based on Erturk and Inman [9, 10]. It describes different parameters that affect the natural frequencies of

PZT bimorphs. On the basis of these parametric studies, one can choose proper geometrical and mechanical constraints to obtain desired natural frequency. The bimorph with a tip mass and an electrical circuit to evacuate power is shown in Figure 5.4.

Assuming damping to be negligible, ω_r the undamped natural frequency of the rth vibration mode in short circuit conditions (i.e., as $R_L \to 0$) is given by:

$$\omega_r = \lambda_r^2 \sqrt{\frac{YL}{mL^4}} \tag{5.2}$$

The eigenvalues of the system (λ_r for rth mode) are determined from following equation:

$$1 + \cos\lambda\cosh\lambda + \lambda\frac{M_t}{mL}\left(\cos\lambda\sinh\lambda - \sin\lambda\cosh\lambda\right) = 0 \tag{5.3}$$

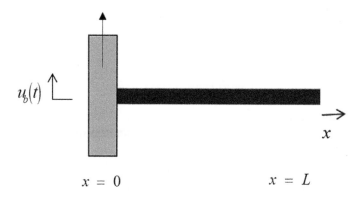

FIGURE 5.3 Cantilever beam excited by the motion to its base in the transverse direction.

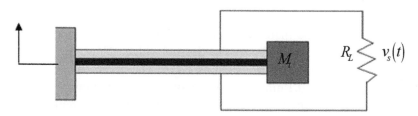

FIGURE 5.4 Bimorph cantilever configurations showing translational base excitation with a series connection.

The natural frequencies of the bimorph with a tip mass can now be calculated from the equation (5.3). The authors considered a smooth radial 195/65R15 91T tire inflated to 220 kPa [2, 3] and found that first mode natural frequency of tire as 515 rad/s. Here a PZT bimorph is designed for this natural frequency. The geometric and material parameters chosen for this design are shown in Tables 5.1 and 5.2.

TABLE 5.1   Geometrical Parameters of Bimorph

Geometric Parameters	Piezoelectric (PZT-5A) (mm)	Substrate (Brass) (mm)
Length, L	20	20
Width, b	3.39	3.39
Thickness, h	0.13	0.08

TABLE 5.2   Material Parameters of Bimorph

Material Parameters	Piezoelectric (PZT-5A)	Substrate (Brass)
Mass density, ρ	7800 kg/m³	9000 kg/m³
Young's modulus, Y	66 GPa	105 GPa
Piezo. Constant, d_{31}	−190 pm/V	—
Permittivity, ε_{33}	1500 ε_0 F/m	—

Brass and PZT-5A are the best options as substrate and piezoelectric material for a bender/bimorph. The bender should fit inside a tiny hollow cylindrical shell, so space constraint for the bender should be considered. We can reduce the length of the bender to make it compact, but the natural frequency will also increase very rapidly; which is not our objective. To reduce the natural frequency we can attach a more weighted tip mass, but it will also increase the volume of tip mass. Due to increased dimensions of tip mass, the bender requires more space for deflection of the bimorph, which will increase the shell size. So the material should be chosen in such a manner that natural frequency should be at the desired value. Thus, the tip mass attached to the bimorph is selected as 1 gm. The given base excitation has an amplitude of 0.1 mm. using the mathematical framework outlined in Erturk and Inman [9–11] expressions for voltage, current, and tip deflection can be derived. The resulting voltage vs. frequency plot is shown in Figure 5.5.

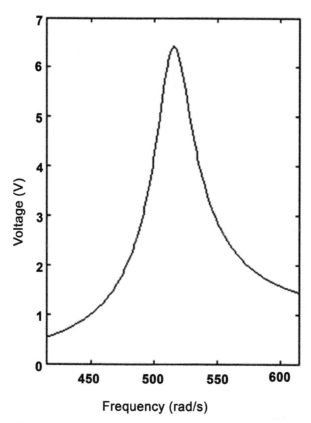

FIGURE 5.5 Voltage vs. frequency of the bimorph at first mode of frequency at 35 $k\Omega$ of load resistance for 195/65R15 91T tire.

The tip deflection is shown in Figure 5.6. Further analysis can be carried out by changing the load resistance and such sensitivity analyses are needed for the final design of the piezo harvester.

5.8 ECONOMIC AND ENVIRONMENTAL ANALYSIS

Safety is the prime motivation for the development of TPMS. The main goal is to reduce the number of accidents occurring every year. The use of TPMS in vehicles can also improve fuel economy and mileage. The US Department of Energy estimated that one psi (pound per square inch) drop in tire pressure is lowering the mileage by 0.4% [12]. This small loss for one vehicle becomes very large when it is multiplied with all running vehicles over a period of time. The estimates show that in the US alone, approximately 100,000,000 cars filling up once per week can save almost 170,000,000 barrels of oil per year by maintaining specified tire pressure. Tests show that visual inspection to judge the specified tire pressure is not reliable.

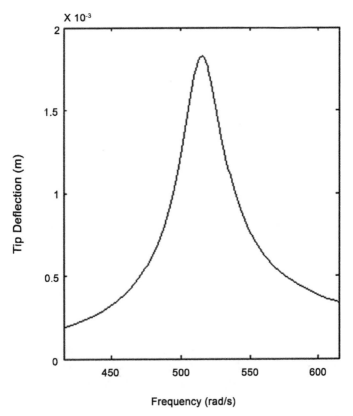

FIGURE 5.6   Tip deflection of the harvester at first mode of frequency at 35 $k\Omega$ of load resistance for 195/65R15 91T tire.

The batteries used in TPMS have a negative environmental impact. The NHTSA estimated a 2% increase in US battery usage. Approximately 71 million tiny batteries are used in the TPMS. But TPMS benefits the environment by saving extra fuel lost due to under-inflation of tire i.e., reduce emission, so a smaller overall increase in battery volume may be justified [4]. Even this battery volume and chemical content can be nullified if we move to piezoelectric harvester operated TPMS. The NHTSA estimates that TPMS saves 660 lives per year, as well as preventing 33,000 injuries and saving $511 million worth of gas [5].

The cost of piezoelectric bimorphs and associated electronics is dropping significantly over the years. As per a roadmap of Freescale the fifth generation TPMS may have energy harvesting capability. Development of a highly miniaturized Tire Pressure Monitoring System for in-tire assembly with volume < 1 cm^3 and mass < 5 gm with MEMS piezo energy harvester is being pursued by a consortium led by Infineon [6]. New piezo materials developed [7, 8, 13] and mass production of piezo devices may make TPMS affordable for all vehicles.

KEYWORDS

- **commercially off the shelf**
- **National Highway Traffic Safety Administration**
- **piezoelectric energy harvester**
- **radio frequency**
- **tire pressure maintenance system**
- **tire pressure monitoring sensor**

REFERENCES

1. Joe, W., (2012). *Tire Pressure Affects Stopping Distance, Fuel Economy.* http://www.newsday.com/classifieds/cars/tire-pressue-affects-stopping-distance-fuel-economy-1.3760110 (accessed on 13 May 2020).
2. Hans, P., (2012). *Tire and Vehicle Dynamic.* SAE International, Elsevier, ISBN: 978–0-08–097016–5.
3. Koizumi, T., Tsujiuchi, N., Matsubara, M., & Nakamura, F., (2010). *Vibration Analysis of Rolling Tire Based on Thin Cylindrical Shell Theory.* Toyo tire and rubber Co., LTD, ISMA.
4. Skip, S., (2011). *Changing TPMS Sensor Batteries.* https://www.tirereview.com/changing-tpms-sensor-batteries/ (accessed on 13 May 2020).
5. Sean, P., (2014). *TPMS Advantages and Disadvantages.* https://www.tirereview.com/author/sean_phillips/ (accessed on 13 May 2020).

6. Thomas, H., et al., (2008). *An Energy Harvesting System for in-tire TPMS.* Collaborative effort between Infineon Austria, Infineon Norway, TU Vienna, Vestfold University College, SINTEF, PwrSoC Workshop.
7. Narayan, K., (2017). *M. Tech Thesis.* IIT Jodhpur, Energy Harvesting through Piezoelectricity and its Application in TPMS.
8. Roger, A., (2009). *Energy Harvesting Looks to Solve Critical TPMS Issues.* http://electronicdesign.com/energy/energy-harvesting-looks-solve-critical-tpms-issues (accessed on 13 May 2020).
9. Alper, E., & Denial, I., (2008). A distributed parameter electromechanical model for cantilevered piezoelectric energy harvesters. *Journal of Vibration and Acoustics.*
10. Alper, E., & Denial, I., (2009). An experimentally validated bimorph cantilever model for piezoelectric energy harvesting from cantilevered beams. *Smart Mater. Struct., 18*(025009), p. 18.
11. Alper, E., & Denial, J. I., (2011). *Piezoelectric Energy Harvesting.* John Wiley & Sons, ISBN: 978-0-470-68254-8.
12. 12. *How Tire Pressure Affects MPG,* (2015). http://procarmechanics.com/how-tire-pressure-affects-mpg/ (accessed on 13 May 2020).
13. 13. Shashank, P., & Inman, D. J., (2009). *Energy Harvesting Technologies.* Springer, ISBN: 978-0-387-76463-4.

CHAPTER 6

A Retrospective Assessment of Elastic-Plastic and Creep Deformation Behavior in Structural Components Like Discs, Cylinders, and Shells

SHIVDEV SHAHI,[1] SATYA BIR SINGH,[2] AND A. K. HAGHI[3]

[1]Research Scholar, Department of Mathematics, Punjabi University, India, E-mail: shivdevshahi93@gmail.com

[2]Professor, Department of Mathematics, Punjabi University, Patiala, India, E-mail: sbsingh69@yahoo.com

[3]Professor Emeritus, Canadian Research and Development Center of Sciences and Cultures, Montreal, Canada

6.1 INTRODUCTION

The significance of studying the deformation behavior of components of complex structures in the form of bars, plates, cylinder, disc etc. is not only essential for the basic understanding of mechanical phenomenon but also to build reliable structures. Advances in the field are targeted to assure safety, reliability, and economy in the design of structure and complete systems, hence to the continued development in design of turbines, aerospace and surface transportation vehicles, prosthetics, material processing, and several other manufacturing technologies. The design engineer is interested in deformation behaviors in connection with his responsibility for avoiding excessive deflection or distortion in machine parts and engineering structures. The mechanical properties of the materials have to be considered for the various parts, thereby optimizing the available resources. The designer is also concerned with identification of the limiting condition of stress which further leads to fracture. Hence, necessary deformation rates have to be analyzed. He may base his calculation as far as they refer to pure elastic deformation upon the theory of elasticity, but he lacks in scope when the plastic state sets in. Therefore, the theory of plasticity is made necessary by demand for better design and more economic use of materials as well as for better control of metal working operations such as wire-drawing and cold rolling. Components such as a disc, cylinder, and a spherical shell will be considered in this chapter and the work done by various researchers will be analyzed to provide insights into the reliability of the structural components.

6.2　STRUCTURAL COMPONENTS IN THE FORM OF DISCS, CYLINDERS, AND SHELLS

Rotating disc play an importance role in machine design. The problems of rotating discs were first treated in the early nineteenth century. Solution of the isotropic discs including variable thickness, variable density, and other case can be found in most of the standard elasticity textbooks [1, 2]. Rotating discs were considered to be one of the exhausted subjects in the field of solid mechanics. Recently new interests have been generated to reinvestigate these problems. The new interest is the result of the development of composite materials and their applications. Composite materials are characterized by high strength to weight ratios, heterogeneity, and anisotropy characteristics. The analytic method has proved to be a simple and reliable means of generating a biaxial state of stress when the loads cannot be directly applied to the material under investigation. In this case, an analytical solution of this problem is needed to interpret the experimental results generated in the laboratory. Because of the steady increase in the use of energy and the impact of that use on the environment, a flywheel made of composite materials has been proved to be an efficient mean of storing energy. Likewise, analytical solution of rotating discs made of composite materials is essential in design and analysis. The calculation of turbine disc involves the solution of second order differential equation having the elastic stress function. This equation is integrable analytically for discs of constant thickness, and for specific examples of discs of non-constant thickness. Arrowsmith [1] has given solutions for discs whose thickness is inversely proportional to radius to the power thickness parameter while Malkin [3] discusses a disc with exponential thickness variation.

A cylinder is a commonly used component in various structural components. In most of the applications, such as pressure vessel for industrial gases, transportation of high-pressurized fluids and piping of nuclear reactors, the cylinder has to operate under severe mechanical and thermal loads, resulting in significant creep and thereby decreasing its service life. Rimrott [4] has solved the problem of thick-walled isotropic cylinders subjected to internal pressure using finite strain theory and has shown that by considering finite strains the creep rate of a thick walled cylinder subject to internal pressure increases even though the creep rate of the same material when subjected to constant true stress in simple tension is constant.

Love [5] studied vibrations and deformation in thin elastic shells. He assumed small deflections and for that, reason neglected all the higher terms of the quadratic in the energy equation. Thus, he obtained linear differential equations for the determination of the equilibrium position of shell under given forces. The buckling of cylindrical shells of uniform thickness under the action of uniformly distributed axial load was further studied but this was till then only confined to monolithic materials. In the design of shells made of functionally graded materials (FGMs) and homogeneous-isotropic materials, it is of technical importance to examine its resistance to bend under expected loading conditions. For that purpose, the determination of the bending load alone is not sufficient in general, but it is further required to clarify the post bending behavior, that is, the behavior of the shell after passing through the buckling load. Firstly, it is necessary to estimate the effect of practically unavoidable imperfections on the buckling load and the second is to evaluate the ultimate strength to exploit the load-carrying capacity of the shell structure. Von Kármán and Tsien [6] argued at the time, geometrical nonlinearity must play an important part in the phenomenon of thin shell deformation. Donnell and Wan [7] reported that the initial geometric imperfection has a significant effect on the buckling and post buckling behavior of cylindrical shells subjected to axial compression. In their analysis, however, the membrane pre-buckling state was assumed, like von Kármán and Tsien [6] did and, therefore, the boundary conditions cannot be incorporated accurately.

Wahl [8, 9] considered a flat disc forged from chromium steel and analyzed it for steady state creep at 538(C. The experimental results for this system were examined theoretically using Tresca and von Mises yield criterion. It is seen that the creep deformations calculated using Tresca criterion provide a better agreement with experimental observations. The results obtained from von Mises criteria are too low in comparison to experimental results, which may be due to the anisotropy of the material used. Author further extended his study to analyze secondary creep in a disc of variable thickness in which he concluded that such a disc had lower difference in maximum and average stress levels as compared to a disc with constant radius.

Rimrott [4] assumed a cylinder made of homogeneous isotropic material with large strain rates and used generally accepted assumptions of constant density, zero axial strain and distortion energy theory and derived equations of creep in a thick walled, closed end, hollow circular cylinder subject to internal pressure considering finite strains, the creep rates increase however the creep rate in the same material subjected to constant true stress in simple tension is constant. Author concluded that thick walled cylinder is an unsafe assumption.

Baker [10] developed a theory for the elastic-plastic response of a thin spherical shell to spherically symmetric internal pressure loading. Analytic solutions were obtained to the linear, small-deflection equations of motion for shell materials which exhibited various degrees of strain-hardening. Numerical solutions obtained by digital computer were also presented for the equations for large deflections obtained by accounting for shell thinning and increase in radius during deformation. The results of this theory are justified by the experimental results.

Ma [11, 12] introduced the parameter of thermal gradient on rotating discs. The analysis was based on the theory of Tresca's yielding criterion and its associated flow rule while the steady state condition was described by power law. The study revealed that the proposed analysis can be used to obtain closed form solutions for creep stresses.

Seth [13] specified that the usage of classical theory of deformation in the transition stage is totally insufficient as it relies on linearity of the stage. Author defined a transition theory which dealt with the non linear character of the deformation and gave a generalized strain measure to study deformation behavior for different kinds of materials under compression. Author argued that the differential system attains some sort of criticality at transition. These critical or transition points can be predicted giving us asymptotic solutions. Author also stated that at the Elastic-plastic deformation stage, the amount of plasticity will rely on geometry of deformation and not on the stresses.

Hulsarkar [14, 40] applied Seth's transition theory of elastic-plastic and creep deformations to solve the problem of creep in composite cylinders subjected to uniform internal pressure and rotating discs with steady state temperature. The generalized expressions for creep transition stresses were obtained, which, in a special case reduces to those derived by assuming the creep laws.

Gupta and Shukla [15] obtained elastics-plastic and transitional stresses in a disc with variable thickness subjected to internal pressure using Seth's transition theory. Non-homogeneity in the disc was due to variation of Poisson's ratio of the material. It was found that the presence of non-homogeneity and thickness variation influence significantly the stresses and pressure required for initial yielding. The thickness variation reduces the magnitude of the stresses and pressure needed for fully-plastic state.

Loghman and Wahab [16] developed a model to estimate creep damages in a thick-walled tube subjected to internal pressure and thermal gradient. The study predicted the changes in creep damage rates during life cycle of the tube due to variation in stresses with time and through-thickness variations. The θ projection concept was used to predict the long-term creep properties up to rupture and the creep rupture data (Figure 6.1).

Güven [17–19] analyzed fully plastic rotating disc and a hyperbolic annular disc with a rigid shaft attached at the bore. The plane state of stress in these conditions has been considered. Exact solutions are obtained using Tresca yielding criterion. The numerical results indicate that maximum stresses and the plastic flow are influenced by the thickness parameter (Figure 6.2).

Singh and Ray [20] first introduced creep analysis in an isotropic functionally graded material rotating disc of Al-SiCp composite. They investigated the creep analysis in an isotropic FGMs rotating disc made of a composite containing silicon carbide particles (SiCp) in a matrix of pure aluminum at uniform elevated temperature by using Norton's power law and they assumed the particle distribution is linearly decreasing from the inner to the outer radius along the radial distance in the disc estimated by regression t of the available experimental data. It is observed that the steady state creep response of the functionally graded material disc in terms of strain rates is significantly superior as compared to a similar disk having uniform distribution of SiCp (Figure 6.3).

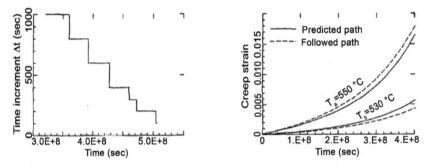

FIGURE 6.1   The time increment variation and predicted path of creep strain [16].

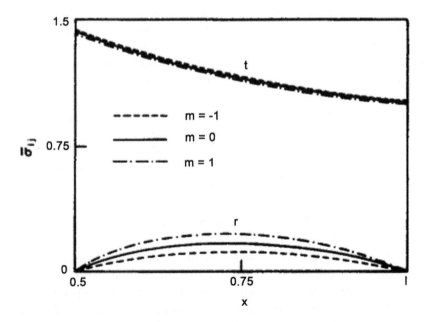

FIGURE 6.2   Stress distribution at fully plastic state with m as thickness parameter [17].

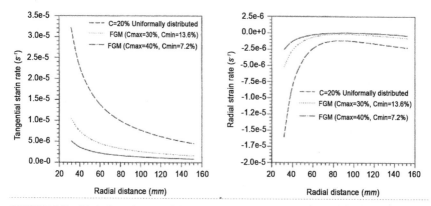

FIGURE 6.3 Variation of tangential and radial strain rate with radial distance in a particle reinforced composite disc with average particle content of 20 vol pct SiC rotating with angular velocity 15,000 rpm at 561 K for the cases when (a) particle content is uniformly distributed in the discs and (b) particle content is linearly decreasing from inner to outer radii. Content is linearly decreasing from inner to outer radii [20].

Singh and Ray [21] extended their work to the analytical treatment of anisotropy and creep in orthotropic aluminum silicon carbide composite rotating disc made of composites containing SiC whiskers under steady state using Hill yield criterion and compared with the results obtained using von Mises yield criterion for the isotropic composites. It is observed that the tangential stress distribution is lower in the middle of the disc but higher near the inner and the outer radius but the radial stress distribution does not get significantly affected due to anisotropy and also observed that anisotropy helped to reduce the tangential strain rate significantly, more near inner radius and the strain rate distribution in the orthotropic disc is lower than that in the isotropic disc following von Mises criterion. It should be noted that the anisotropy constants taken from the experimental results of other studies and the lowering of tangential creep rate may be significant in the context of real life engineering. The compressive radial strain rate also reduced in the disc following Hill criterion of yield plasticity as compared to that in isotropic disc. Thus, anisotropy appeared to help in restraining creep response both in the tangential and in the radial directions (Figure 6.4).

Eraslan and Akis [22] obtained the plane strain analytical solutions, based on Tresca's yield criterion, for FG elastic and elastic-plastic pressurized tube problems using small deformation theory. The Young's modulus and the uniaxial yield limit of the tube material were assumed to vary radially according to two-parametric parabolic forms. The study also investigated various types of stress states i.e., elastic, partially plastic and fully plastic. It is observed that the elasto-plastic response of the FG pressurized tube has been significantly affected by the material nonhomogenity. The non-homogeneous elasto-plastic solution has also been observed to reduce to that obtained for homogeneous one, when the material parameters were appropriately selected.

You et al. [23] analyzed steady state creep in thick-walled cylinders made of arbitrary FGM and subjected to internal pressure. The stresses and strain rates were calculated by using Norton's creep law. The impact of radial variations of material parameters was investigated on stresses in the cylinder.

Abrinia et al. [24] obtained analytical solution to obtain radial and circumferential stresses in a functionally graded thick cylindrical vessel under the influence of internal pressure and temperature. The effect of non-homogeneity in functionally graded cylinder was analyzed in the context of achieving the lowest stress levels in the cylinder (Figure 6.5).

Thakur [25–27] have studied elastic plastic and creep transition comprehensively in structures such as rotating annular and solid discs, circular cylinders with and without inclusion, cylindrical, and spherical shells subject to various parameters such as angular speed, variable density, thermal gradients and edge loading using Seth's transition theory. Variations in radial and circumferential stresses have been determined for different situations for compressible and incompressible materials and results for situations of fractures in subsequent situations have been inferred.

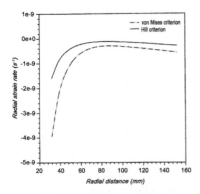 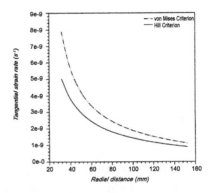

FIGURE 6.4 Variation of radial and tangential strain rates with radial distance in a rotating disc yielding according to von Mises isotropic yield criterion and Hill orthotropic yield criterion [21].

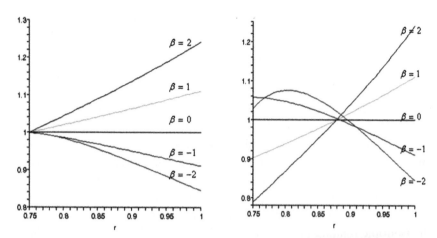

FIGURE 6.5 Variation in radial and circumferential stresses in a cylindrical vessel under the combined loading of pressure and temperature [24].

Rattan, Chamoli, Singh, and Nishi [28] studied creep behavior of anisotropic functionally graded rotating discs. They investigated the creep behavior of an anisotropic rotating disc of functionally gradient material using Hill's yield criteria by Sherby's constitutive model. It is concluded that the anisotropy of the material has a significant effect on the creep behavior of the FGM disc. It is also observed that the FGM disc shows better creep behavior than the non FGM disc.

Vandana and Singh [29] analyzed the effect of residual stress and reinforcement geometry in an anisotropic composite rotating disc having varying thickness. The creep analysis carried out in a rotating disc made of Al-SiC (particle/whisker) composite having hyperbolically varying thickness using anisotropic Hoffman yield criterion and results obtained are compared with those using Hill's criterion ignoring difference in yield stresses by Sherby's creep law for the steady state creep behavior. It is observed that the stresses are not much affected by the presence of thermal residual stress, while thermal residual stress introduces significant change in the strain rates in an anisotropic rotating disc. Secondly, it is noticed that the steady state creep rates in whisker reinforced disc with/without residual stress are observed to be significantly lower than those observed in particle reinforced disc with/without residual stress. It is concluded that the presence of residual stress in an anisotropic disc with varying thickness needs attention for designing a disc.

Nejad et al. [41] studied a closed-form analytical solution for the steady-state creep stresses of rotating thick cylindrical pressure vessels made of FGMs. Creep response of the material has been described by Norton's Law. Exact solutions for stresses and strain rate were obtained under the plane strain condition. The effect on radial and circumferential stresses together with the equivalent strain rate in rotating thick-walled cylindrical vessels under internal pressure has been investigated. The result obtained shows that the property of FGMs has a significant influence on the equivalent creep strain rate and stresses distributions along the radial direction (Figure 6.6).

Vandana and Singh [30] studied mathematical modeling of creep in a functionally graded rotating disc with varying thickness. This study gave an analytical framework for the analysis of creep stresses and creep rates in the isotropic rotating non-FGM/FGM disc with uniform and varying thickness by Sherby's law. The creep response of rotating disc is expressed by threshold stress with value of stress exponent as 8. The results compared for isotropic non FGM/FGM constant thickness disc with those estimated for isotropic varying thickness disc with the same average particle content distributed uniformly and suggested the distribution of stresses and strain rates becomes relatively more uniform in the isotropic FGM hyperbolic thickness disc (Figure 6.7).

Loghman et al. [31] investigated the creep behavior of composite cylinder made of polypropylene reinforced by functionally graded multi walled carbon nanotubes (FG-MWCNTs). It had been found that the radial and circumferential strains of the cylinder reduce with increasing content of carbon nanotubes. It was also concluded that the uniform distribution of MWCNTs reinforcement does not considerably influence on stresses.

Sharma and Panchal [32] considered a pressurized thick-walled functionally graded rotating spherical shell. Creep stresses were calculated under thermal effects. Transition theory and the concept of generalized measure were used to investigate these creep stresses. It was observed that angular velocity and non-homogeneity plays significant role on thermal creep stresses. Further, it was concluded that spherical shell made of thick walls is a safer prospect for designs.

Thakur et al. [33] presented exact solution of rotating disc with shaft problem in the elastoplasic state of stress having variable density and thickness by using Seth's transition theory. It is observed that the rotating disc made of the compressible material with an inclusion requires higher angular speed to yield at the internal surface as compared to the disc made of incompressible material, and a much higher angular speed is required to yield with the increase in radii ratio. The thickness and density parameters decrease the value of angular speed at the internal surface of the rotating disc of compressible as well as incompressible materials.

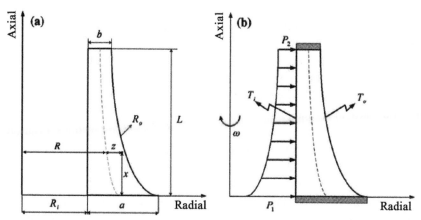

FIGURE 6.6 Cross section of thick cylindrical vessel with variable thickness [35].

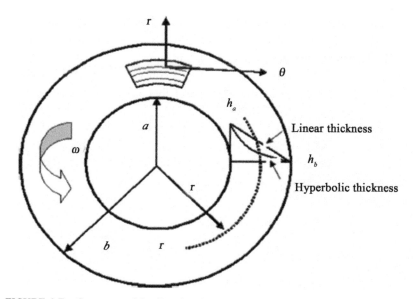

FIGURE 6.7 Geometry of the disc showing variation in thickness [30].

6.3 DISCUSSIONS AND CONCLUSION

The linear theory has given rise to a vast amount of important results which have been corroborated by experiments more exactly than one may expect, covering all types of structural components. Elastic-plastic transition is obtained in current literature with the help of a semi-empirical yield condition like that of Tresca or Von-Mises. The stresses are obtained from the elastic solution and then substituted in the yield condition to get the transition surface. The possibility of treating it as a transition or turning point phenomenon in finite deformation has not been explored. When the plastic state tends to set in, the stress-strain relation undergoes a change. This is reflected in the equations. The linear classical theory cannot do it. In the transition state whole of the material participates, and not simply a selected region or a line as assumed by classical theories. The demands of high speed technology in transportation, communication, and energy conversion have forced us to take serious notice of non-linearity. Since this non-linearity is difficult to investigate, workers have taken to the artifice of replacing it by singular, non-differentiable, or discontinuous surfaces. This piecewise treatment necessitates the use of adhoc and semi-empirical law, which may or may not exist. Linearizing non-linear problems by perturbation, boundary layer and other techniques do not provide satisfactory explanation for some important characteristics of non-linearity. As a result, a number of important physical effects do not get adequate scientific expression. There is hardly any information which we do not know about linear fields. Their existence, uniqueness, and stability are well established. Nature does not always confirm to our abstract concepts of linearity, smoothness, symmetry, identity, and isotropy. It is true that physical phenomena tend to behave linearly in course of time, which may be millions of years. Thus transitions, which frequently occur in nature, have to be tackled. In fact, all inter-disciplinary fields which are so important in modern research give rise to important transition problems. Hence, both the macro and micro analysts must devote attention to non-linearity involved in the subject.

KEYWORDS

- **creep**
- **functionally graded materials**
- **functionally graded multi walled carbon nanotubes**

- **isotropic cylinders**
- **non-linear problems**
- **perturbation**
- **semi-empirical law**

REFERENCES

1. Arrowsmith, G., (1923). *The Design of Rotating Discs* (Vol. 116, p. 417.). Engineering, London.
2. Heyman, J., (1958). Plastic design of rotating discs. *Proc, Inst. Mech. Engrs., 172,* 531–546.
3. Malkin, I., (1934). Design and calculation of steam turbine disk wheels. *Trans. AMER. Soc. Mech. Engrs., 56,* 585.
4. Rimrott, F. P. J., (1959). Creep of thick-walled tube under internal pressure considering large strains. *Journal of Applied Mechanics-Transactions ASME, 26,* 271–274.
5. Love, A. E. H., (1888). The small free vibrations and deformation of a thin elastic shell. *Phil. Trans. R. Soc., 179A,* 491–546.
6. Von, K. T., & Tsien, H. S., (1941). The buckling of thin cylindrical shells under axial compression. *Journal of the Aerospace Sciences, 8,* 303–312.
7. Donnell, L. H., & Wan, C. C., (1950). Effect of imperfections on buckling of thin cylinders and columns under axial compression. *Journal of Applied Mechanics ASME, 17,* 73–83.
8. Wahl, A. M., Sankey, G. O., Manjoine, M. J., & Shoemaker, E., (1954). Creep tests of rotating disks at elevated temperature and comparison with theory. *Journal of Applied Mechanics, 76,* 225–235.
9. Wahl, A. M., (1957). Stress distribution in rotating discs subjected to creep at elevated temperature. *Journal of Applied Mechanics, ASME Transactions, 29,* 299–305.
10. Baker, W. E., (1960). The elastic-plastic response of thin spherical shells to internal blast loading. *J. Appl. Mech., 27*(1), 139–144.
11. Ma, B. M., (1961). Creep analysis of rotating solid disks with variable thickness and temperature. *Jour. Franklin Inst., 271,* 40–55.
12. Ma, B. M., (1964). A power function creep analysis for rotating solid disks having variable thickness and temperature. *Journal of the Franklin Institute, 277*(6), 593–612.
13. Seth, B. R., (1962a). Transition theory of elastic-plastic deformation, creep, and relaxation. *Nature, 195,* 896–897,
14. Suresh, H., (1979). Thermo elastic-plastic transition in rotating disks with steady state temperature. *Indian J. Pure Appl. Math., 10*(5), 567–601.
15. Gupta, S. K., & Shukla, R. K., (1994). Effect of non-homogeneity on elastic-plastic transition in a thin rotating disk. *Indian J. Pure Appl. Math., 25*(10), 1089–1097.
16. Loghman, A., & Wahab, A., (1996). Creep damage simulation of thick-walled tubes using the θ projection concept. *Int. J. of Pressure Vessels and Piping, 67*(1), 105–111.

17. Güven, U., (1997). The fully plastic rotating disk with rigid inclusion. *ZAMM, 77*(9), 714–716.
18. Güven, U., (1998). Elastic-plastic stress distribution in a rotating hyperbolic disk with rigid inclusion. *International Journal of Mechanical Sciences, 40(1)*, 97–109.
19. Güven, U., (1999). Elastic-plastic rotating disk with rigid inclusion. *Mechanics of Structures and Machines, 27*(1), 117–128.
20. Singh, S. B., & Ray, S., (2001). Steady-state creep behavior in an isotropic functionally graded material rotating disc of Al-SIC composite. *Metallurgical Transactions, 32*, 1679–1685.
21. Singh, S. B., & Ray, S., (2002). Modeling the anisotropy and creep in orthotropic aluminum-silicon carbide composite rotating disc. *Mechanics of Materials, 34*(6), 363–372.
22. Eraslan, A. N., & Akis, T., (2006). Plane strain analytical solution for a functionally graded elastic-plastic pressurized tube. *Int. J. of Pressure Vessels and Piping, 83*, 635–644.
23. You, L. H., Ou, H., & Zheng, Z. Y., (2007). Creep deformations and stresses in thick-walled cylindrical vessels of functionally graded materials subjected to internal pressure. *Composite Structures, 78*, 285–291.
24. Abrinia, K., Naee, H., Sadeghi, F., & Djavanroodi, F., (2008). New analysis for the FGM thick cylinders under combined pressure and temperature loading. *American J. of Applied Sci., 5*(7), 852–859.
25. Pankaj, T., (2010). Elastic-plastic transition stresses in a thin rotating disc with rigid inclusion by infinitesimal deformation under steady-state temperature. *Thermal Science, 14*(1), 209–219.
26. Pankaj, T., (2011). Effect of transition stresses in a disc having variable thickness and poison's ratio subjected to internal pressure. *WSEAS Transactions on Applied and Theoretical Mechanics, 6*(4), 147–159.
27. Pankaj, T., (2012). Elastic-plastic transitional stresses in a thin rotating disk with loading edge stresses in a thin rotating disc of variable thickness with rigid shaft. *Journal of Technology for Plasticity, Serbia, 37*(1), 1–14.
28. Neeraj, C., Minto, R., Singh, S. B., & Nishi, G., (2013). Creep behavior of anisotropic functionally graded rotating discs. *International Journal of Computational Materials Science and Engineering, 12*(1). Imperial College Press.
29. Vandana, G., & Singh, S. B., (2013). Effect of residual stress and reinforcement geometry in an anisotropic composite rotating disc having varying thickness. *International Journal of Computational Materials Science and Engineering.*
30. Vandana, G., & Singh, S. B., (2016). Mathematical modeling of creep in a functionally graded rotating disc with varying thickness. *The Regenerative Engineering Society Regen. Eng. Transl. Med.* doi: 10.1007/s40883–016–0018–3.
31. Loghman, A., Shayestemoghadam, H., & Loghman, E., (2016). Creep evolution analysis of composite cylinder made of polypropylene reinforced by functionally graded MWCNTs. *Journal of Solid Mechanics, Article 11, 8*(2), 372–383.
32. Sanjeev, S., & Rekha, P., (2017). Thermal creep deformation in pressurized thick-walled functionally graded rotating spherical shell. *International Journal of Pure and Applied Mathematics, 114*(3), 435–444.

33. Pankaj, T., Monika, S., Shivdev, S., Singh, S. B., Fadugba, S. E., & Jasmina, L. S., (2018). Modeling of creep behavior of a rotating disc in the presence of load and variable thickness by using Seth transition theory. *Structural Integrity and Life, 18*(2), 133–140.

34. Neeraj, C., Minto, R., & Singh, S. B., (2010). Effect of anisotropy on the creep of a rotating disc of Al-SiCp composite. *International Journal of Contemporary Mathematical Sciences, 5*(11), 509–516.

35. Nejad, M. Z., Abedi, M., Lotfian, M. H., & Ghannad, M., (2013). Elastic analysis of exponential FGM vessel subjected to internal and external pressure. *Central European Journal of Engineering, 3*, 459–465.

36. Pankaj, T., (2010). Creep transition stresses in a thin rotating disc with shaft by finite deformation under steady-state temperature. *Thermal Science, 14*(2), 425–436.

37. Pankaj, T., Sethi, M., Shivdev, S., Singh, S. B., & Emmanuel, F. S., (2018). Exact solution of rotating disc with shaft problem in the elastoplastic state of stress having variable density and thickness. *Structural Integrity and Life, 18*(2), 126–132.

38. Seth, B. R., (1962). The general measure in deformation. *Proc. IUTAM on Second Order Effects in Elasticity, Plasticity and Fluid Dynamics, Haifa.*

39. Singh, S. B., & Ray, S., (1998). Influence of anisotropy on creep in a whisker reinforced MMC rotating disc. *Proceeding of the National Seminar on Composite Materials: COMPEAT-98* (pp. 83–102).

40. Hulsarkar, S. (1981). Elastic plastic transitions in transversely isotropic shells under uniform pressure. *Indian J. Pure Applied Math. 12*(4), 552–557.

41. Nejad, M., Hoseini, Z., Niknejad, A., & Ghannad, M. (2015). Steady-State Creep Deformations and Stresses in FGM Rotating Thick Cylindrical Pressure Vessels. *Journal of Mechanics, 31*(1), 16.

CHAPTER 7

Investigation of Elastic-Plastic Transitional Stresses in Zirconia-Based Ceramic Dental Implants under Uniaxial Compression

SHIVDEV SHAHI,[1] SATYA BIR SINGH,[2] and A. K. HAGHI[3]

[1]*Research Scholar, Department of Mathematics, Punjabi University, India, E-mail: shivdevshahi93@gmail.com*

[2]*Professor, Department of Mathematics, Punjabi University, Patiala, India, E-mail: sbsingh69@yahoo.com*

[3]*Professor Emeritus, Canadian Research and Development Center of Sciences and Cultures, Montreal, Canada*

7.1 INTRODUCTION

Zirconia (ZrO_2) has been considered as a wear-resistant and osteoconductive ceramic, highly suitable for making stress-bearing crowns of dental implants. Various properties, namely, less inflammatory response, compatibility of zirconia in contact with gum tissues, lower plaque retention, and highly aesthetic tooth-like coloration has made zirconia a suitable alternative to titanium implants [2, 6, 19]. The production of ceramic dental implants have improved in past few years [7, 20]. The yttria-stabilized tetragonal zirconia polycrystal, popularly known as Y-TZP has been a preference for some years now, having the highest fracture strength [10, 12, 15, 27]. Muhammad et al. [11] confirmed in their chapter that zirconia is highly flexible in all directions of the lattice plane. The Poisson ratio obtained, i.e., from 0.16 to 0.31 also indicate high anisotropy. The stiffness constants so obtained will also be used in this chapter for the determination of elastic-plastic transitional stresses.

This chapter is concerned with the investigation of elastic-plastic transition stress in Zirconia-based crowns of ceramic dental implants, further
comparing the results with stresses in titanium-based implants. The chapter
also checks the similarities between the implants and human tooth enamel
using Seth's transition theory. The crowns will be modeled as spherical
shells exhibiting transversely isotropic macrostructural symmetry. Equations for modeling spherical shells made of isotropic materials are available
in most standard textbooks [3, 9, 13, 14, 17, 29]. Following the classical
methods of stress and strain determination, Miller [16] evaluated solutions
for stresses and displacements in a thick spherical shell subjected to internal
and external pressure loads. You et al. [30] presented a highly precise model
to carry out elastic analysis of thick-walled spherical pressure vessels. The
authors have studied the behavior of shells particularly when some assumptions, such as (i) incompressibility of material used, (ii) creep strain law
derived by Norton, (iii) yield condition of Tresca, and (iv) associated flow
rules; were made. The need of utilization of these specially appointed semi-
experimental laws in elastic-plastic transition depends on approach that the
transition is a linear phenomenon which is unrealistic. Deformation fields
related with the irreversible phenomenon, such as elastic-plastic disfigure-
ments, creep relaxation, fatigue, and crack, etc. are non-linear in character.
The traditional measures of deformation are not adequate to manage transitions. The concept of generalized strain measures and transition theory
given by Seth [21] has been applied to find elastic-plastic stresses in various
problems by solving the non-linear differential equations at the transition
points. Shahi and Singh [24] successfully calculated elastic-plastic transitional stresses in human tooth enamel and dentine using this theory. The
results for hydroxyapatite (HAP)-$Ca_{10}(PO_4)_6(OH)_2$, which structures 95% of
enamel and 50% of dentine by weight, were also obtained for comparative
analysis. The theory has been used to solve various problems of stress and
strain determination in structures modeled in the form of discs and shells
[25, 28]. All these problems based on the recognition of the transition state
as separate state necessitates showing the existence of the used constitutive
equation for that state.

7.2   GOVERNING EQUATIONS

We consider a spherical shell of constant thickness with internal and
external radius a and b respectively under external pressure p. The

external pressure will act radially to simulate the state of axial compression (Figure 7.1).

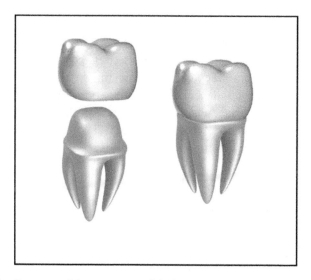

FIGURE 7.1   Structure of the crown part of the implant.

1. **Displacement Coordinates:** The components of displacement in spherical coordinates (r, θ, ϕ) are taken as:

$$u = r(1-\beta); \; v = 0; \; w = 0 \tag{7.1}$$

where: β is position function depending on r.
The generalized components of strain are given by Seth [21, 22] as:

$$e_{rr} = \frac{1}{n}\left[1 - \left(r\beta' + \beta\right)^n\right] = \frac{1}{n}\left[1 - \beta^n\left(1 + P\right)^n\right]$$

$$e_{\theta\theta} = e_{\phi\phi} = \frac{1}{n}\left[1 - \beta^n\right] \tag{7.2}$$

$$er\theta = e\theta\phi = e\phi r = 0$$

where: n is the measure and $r\beta' = \beta P$; P is a function of β and β is a function of r.

2. **Stress-Strain Relation:** The stress-strain relations for isotropic material are given by Sokolinokoff [26]:

$$T_{ij} = c_{ijkl}\, e_{kl}, \ (i, j, k, l = 1, 2, 3)$$

where T_{ij} and e_{kl} are the stress and strain tensors respectively. These nine equations contain a total of 81 coefficients e_{ijkl}, but not all the coefficients are independent. The symmetry of T_{ij} and e_{ij} reduces the number of independent coefficients to 36. For Transversely isotropic materials which have a plane of elastic symmetry, these independent coefficients reduce to 5. The constitutive equations for transversely isotropic media are given by Altenbach et al. [1]:

$$\begin{bmatrix} T_{11} \\ T_{22} \\ T_{33} \\ T_{23} \\ T_{31} \\ T_{12} \end{bmatrix} = \begin{bmatrix} c_{11} & c_{12} & c_{13} & 0 & 0 & 0 \\ c_{12} & c_{11} & c_{13} & 0 & 0 & 0 \\ c_{13} & c_{13} & c_{33} & 0 & 0 & 0 \\ 0 & 0 & 0 & c_{44} & 0 & 0 \\ 0 & 0 & 0 & 0 & c_{44} & 0 \\ 0 & 0 & 0 & 0 & 0 & \frac{1}{2}(c_{11}-c_{12}) \end{bmatrix} \begin{bmatrix} e_{11} \\ e_{22} \\ e_{33} \\ e_{23} \\ e_{31} \\ e_{12} \end{bmatrix}$$

(7.3)

Substituting Eqn. (7.2) in Eqn. (7.3), we get:

$$T_{rr} = \frac{c_{33}}{n}\left[1-\left(r\beta'+\beta\right)^{n}\right]+\frac{1}{n}\left(2c_{12}\right)\left(1-\beta^{n}\right) \Rightarrow c_{33}e_{rr}$$

(7.4)

$$T_{\theta\theta} = T_{\varphi\varphi} = \frac{c_{21}}{n}\left[1-\left(r\beta'+\beta\right)^{n}\right]+\frac{2}{n}\left(c_{11}-c_{66}\right)\left(1-\beta^{n}\right) \Rightarrow c_{12}e_{rr}+2\left(c_{11}-c_{66}\right)e_{\theta\theta};$$

$$T_{r\theta} = T_{\theta\phi} = T_{\phi r} = 0$$

3. **Equation of Equilibrium:** The equations of equilibrium are:

$$\frac{\partial T_{rr}}{\partial r}+\frac{1}{r\sin\theta}\frac{\partial T_{r\phi}}{\partial \phi}+\frac{1}{r}\frac{\partial T_{r\theta}}{\partial \theta}+\frac{2T_{rr}-T_{\theta\theta}-T_{\phi\phi}+T_{r\theta}\cot\theta}{r}=0$$

;

;

$$\frac{\partial T_{r\theta}}{\partial r} + \frac{1}{r\sin\theta}\frac{\partial T_{\theta\phi}}{\partial\phi} + \frac{1}{r}\frac{\partial T_{\theta\theta}}{\partial\theta} + \frac{3T_{r\theta} + \left(T_{\theta\theta} - T_{\phi\phi}\right)\cot\theta}{r} = 0$$
$$;\quad (7.5)$$

$$\frac{\partial T_{r\phi}}{\partial r} + \frac{1}{r\sin\theta}\frac{\partial T_{\phi\phi}}{\partial\phi} + \frac{1}{r}\frac{\partial T_{\phi\theta}}{\partial\theta} + \frac{3T_{r\phi} + 2T_{\theta\theta}\cot\theta}{r} = 0.$$

Substituting Eqn. (7.4) in Eqn. (7.5), we see that the equations of equilibrium are all satisfied except:

$$\frac{\partial T_{rr}}{\partial r} + \frac{2T_{rr} - T_{\theta\theta} - T_{\phi\phi}}{r} = 0 \qquad (7.6)$$

$$or\ \frac{\partial T_{rr}}{\partial r} + \frac{2}{r}(T_{rr} - T_{\theta\theta}) = 0 \qquad (7.7)$$

From Eqn. (7.7), one may also say that:

$$T_{\phi\phi} - T_{\theta\theta} = 0 \qquad (7.8)$$

Eqn. (7.8) is satisfied by $T_{\theta\theta}$ and $T_{\phi\phi}$ as given by Eqn. (7.2). If $c_{21} = c_{31}$, $c_{22} - c_{33} - c_{32} - c_{23}$ the equation of equilibrium from Eqn. (7.6) becomes:

$$\frac{\partial T_{rr}}{\partial r} + \frac{2\left(T_{rr} - T_{\theta\theta}\right)}{r} = 0, \qquad (7.9)$$

4. **Critical Points or Turning Points:** By substituting Eqn. (7.4) into Eqn. (7.9), we get a non-linear differential equation in terms of β:

$$P(P+1)^{n+1}\beta\frac{dP}{d\beta} + P(P+1)^{n} + 2(1-C_{1})P - \frac{2}{n\beta^{n}}[Q_{1} - Q_{2}] \quad (7.10)$$

where;

$Q_{1} = C_{1}\{1-\beta^{n}(1-P)^{n}\}$, $Q2 = C2(1-C1)\{1-\beta^{n}\}$, $C_{1} = (c_{33} - c_{13})/c_{33}$

and

$$C_2 = (c_{11} + c_{12} - 2c_{13}) / (c_{33}(1-C_1))$$

where P is function of β and β is function of r only.

5. **Transition Points:** The transition points of β in Eqn. (7.10) are $P = 0$, $P \to -1$ and $P \to \pm\infty$

    To solve the Elastoplastic stress problems we consider the case of $P \to \pm\infty$

6. **Boundary Condition:** The boundary conditions of the problem are given by:

$$r = a, \ \tau_{rr} = 0$$
$$r = b, \ \tau_{rr} = -P \qquad\qquad (7.11)$$

7.3   PROBLEM SOLUTION

For finding the elastic-plastic stresses, the transition function is taken through the principal stresses at the transition point $P \to \pm\infty$, we define the transition function ζ as:

$$\zeta = (3 - 2C_1) - \frac{nT_{rr}}{(c_{33})} \cong \left[\beta^n (P+1)^n + 2(1 - C_1) \right] \qquad (7.12)$$

where ζ be the transition function unction of r only. Taking the logarithmic differentiation of Eqn. (7.12), with respect to r and using Eqn. (7.10), we get:

$$\frac{d(\log \zeta)}{dr} = \frac{-2C_1}{r} \qquad\qquad (7.13)$$

Taking the asymptotic value of Eqn. (7.13) as $P \to \pm\infty$ and integrating, we get:

$$\zeta = Ar^{-2C_1} \qquad\qquad (7.14)$$

where A is a constant of integration and $C_1 = (c_{33} - c_{13})/c_{33}$. From Eqn. (7.12) and (7.14), we have:

$$T_{rr} = \frac{c_{33}}{n}\left[(3-2C_1) + Ar^{-2C_1}\right] \tag{7.15}$$

Using boundary condition from Eqn. (7.11) in Eqn. (7.15), we get:

$$A = -\frac{(3-2C_1)}{a^{-2C_1}} \quad \text{and} \quad p = -\frac{c_{33}}{n}\left[(3-2C_1)\left[1-\left(\frac{b}{a}\right)^{-2C_1}\right]\right] \tag{7.16}$$

Substituting Eqn. (7.16) in to Eqn. (7.15) and using Eqn. (7.16) in equation of equilibrium, we get:

$$T_{rr} = \frac{c_{33}}{n}\left[(3-2C_1)\left[1-\left(\frac{r}{a}\right)^{-2C_1}\right]\right]$$

$$T_{\theta\theta} = T_{rr} + \frac{c_{33}}{n}\left[(3-2C_1)C_1\left(\frac{r}{a}\right)^{-2C_1}\right] \tag{7.17}$$

1. **Initial Yielding:** From Eqn. (7.17), it is seen that $|T_{\theta\theta} - Trr|$ is maximum at the outer surface (that is at $r = b$), therefore yielding of the shell will take place at the external surface of the shell:

$$\left|T_{\theta\theta} - T_{rr}\right|_{r=b(external\,surface)} = \left|\frac{c_{33}}{n}\left[(3-2C_1)C_1\left(\frac{b}{a}\right)^{-2C_1}\right]\right| \equiv Y\,(Yielding) \tag{7.18}$$

Using Eqn. (7.18) in Eqns. (7.15)–(7.17), we get the transitional stresses as in non-dimensional components as:

$$\sigma_{rr} = \frac{1}{C_1}\frac{\left(1-R^{-2C_1}\right)}{\left(R_0^{-2C_1}\right)}; \quad \sigma_{\theta\theta} = \frac{1}{C_1}\frac{\left(1-R^{-2C_1}\right)}{\left(R_0^{-2C_1}\right)} + \left(\frac{R^{-2C_1}}{R_0^{-2C_1}}\right) \quad P_{oi} = \frac{\left(R_0^{-2C_1}-1\right)}{C_1 R_0^{-2C_1}} \quad \text{and}$$

$$\tag{7.19}$$

where;

$$R = r / a, \, R_0 = b / a, \, \sigma_{rr} = T_{rr} / Y, \, \sigma_{\theta\theta} = T_{\theta\theta} / Y$$

and

$$P_{oi} = p/Y$$

2. **Fully-Plastic State:** For fully-plastic case [23], $C_1 \rightarrow 0$; therefore stresses and pressure from Eqn. (7.19) becomes:

$$\sigma_{rr} = 2Y^* \log R_0, \, \sigma_{\theta\theta} = 2Y^* \log R \text{ and } P_{of} = Y^* + \sigma_{rr} \qquad (7.20)$$

where,

$$R = r/a, \, R_0 = b/a, \, \sigma_{rr} = T_{rr}/Y^*, \, \sigma_{\theta\theta} = T_{\theta\theta}/Y^*$$

and

$$P_{of} = p/Y^*$$

7.4   NUMERICAL RESULTS AND DISCUSSION

The above investigations elaborate the initial yielding and fully plastic state of a crown made of zirconia and titanium modeled in the form of spherical shell subjected to external pressure, to analyze uniaxial compression. The elastic constants for the same are taken from the literature [11] which has been obtained by ultrasonic resonance spectroscopy, a nondestructive measure to obtain the stiffness constants. The results obtained for both types of crowns are compared with enamel. Enamel is made up of HAP mineral, 95% by vol. All these materials exhibit transversely isotropic macrostructural symmetry.

In Figure 7.2, the curves are plotted for pressure at initial yielding at various radius ratios. It is observed that the intensity of pressure at initial yielding increases with an increase in the thickness of the shell. It has been observed that titanium had the lowest yield strength and yielded at lower levels of stress as compared to enamel and Zirconia. Figures 7.3 and 7.4 show the trends of radial and circumferential stresses at initial yielding. Maximum stresses were observed at the external surface of the shell. In Figure 7.5, the curves are plotted for the pressure required at a fully plastic state for various radius ratios. It has been observed that shells exhibited high plasticity when the thickness of the shell was between ratios $1 < R_0 < 3$, particularly Zirconia. A significant drop in the levels of plasticity is observed when the thickness increases. The plasticity of Zirconia is inferred to be greater than that of titanium. Figures 7.6 and 7.7 represent the trends of radial and circumferential stresses at a fully plastic state. The observations infer to the fact that the principal stress differences were maximum at the external surface of the crowns.

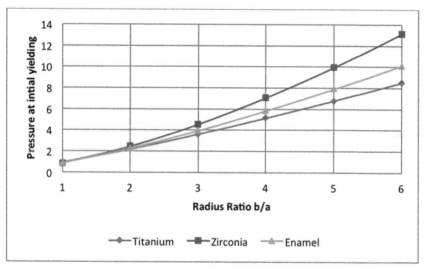

FIGURE 7.2 Pressure in the Shell at initial yielding for radius ratio R_0.

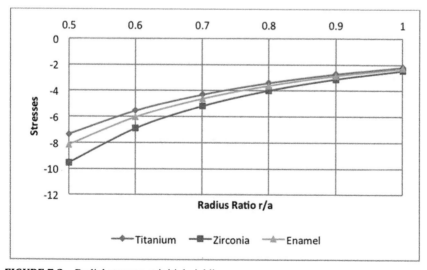

FIGURE 7.3 Radial stresses at initial yielding.

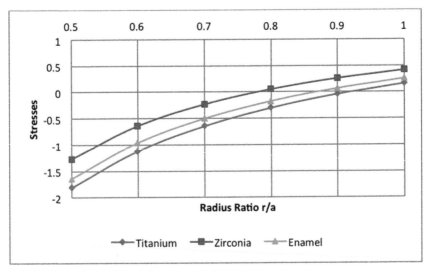

FIGURE 7.4 Circumferential stresses at initial yielding.

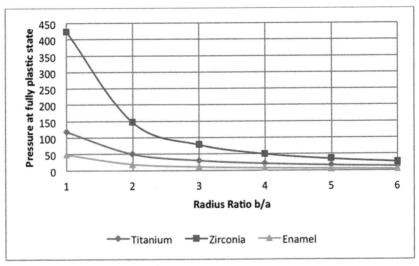

FIGURE 7.5 Pressure in the shell at fully plastic state for radius ratio R_0.

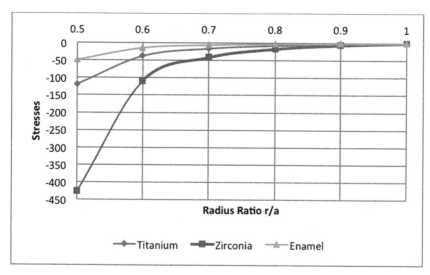

FIGURE 7.6   Radial stresses at fully plastic state.

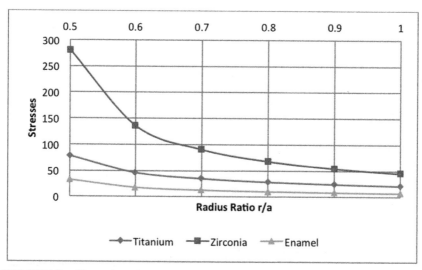

FIGURE 7.7   Circumferential stresses at fully plastic state.

7.5   CONCLUSIONS

The findings allow us to conclude that zirconia has a greater resemblance with enamel with the necessary elastic and plastic limit, which demonstrates considerable ability to suppress a crack growth. Varying values of pressure required for initial yielding and fully plastic state were calculated for various radius ratios depending on the geometry of the crown sample. Trends of the graphs were similar for enamel and HAP due to enamel's composition. The significant difference between stress buildup at the inner and outer layer of the implant crown is observed by varying the radii ratios.

KEYWORDS

- **boundary condition**
- **critical points**
- **equilibrium**
- **hydroxyapatite**
- **initial yielding**
- **stress-strain relation**

REFERENCES

1. Altenbach, H., Altenbach, J., & Kissing, W., (2004). *Mechanics of Composite Structural Elements*. Springer-Verlag.
2. Bianchi, A. E., Bosetti, M., Dolci, G., Sberna, M. T., Sanfilippo, F., & Cannas, M., (2004). *In vitro* and *in vivo* follow-up of titanium transmucosal implants with zirconia collar. *J. Appl. Biomater. Biomech., 2*, 143–150.
3. Boyle, J. T., & Spence, J., (1983). *Stress Analysis for Creep*. Butterworth's Coy. Ltd. London.
4. Cuy, J. L., Mann, A. B., Livi, K. J., Teaford, M. F., & Weihs, T. P., (2002). Nanoindentation mapping of the mechanical properties of human molar tooth enamel. *Arch. Oral. Biol., 47*, 281–291.
5. Zaytsev, D., Grigoriev, S., & Panfilov, P., (2012). Deformation behavior of human dentin under uniaxial compression. *Int. J. Biomater, 8*. Article ID: 854539.
6. Degidi, M., Artese, L., Scarano, A., Perrotti, V., Gehrke, P., & Piattelli, (2006). A. Inflammatory infiltrate, microvessel density, nitric oxide synthase expression, and

proliferative activity in peri-implant soft tissues around titanium oxide healing caps. *J. Periodontol, 77*, 73–80.

7. Depprich, R., Ommerborn, M., Zipprich, H., Naujoks, C., Handschel, J., & Wiesmann, H. P., (2008). Behavior of osteoblastic cells cultured on titanium and structured zirconia surfaces. *Head Face Med., 4*, 29.

8. Menéndez-Proupina, E., Cervantes-Rodríguezb, S., Osorio-Pulgara, R., Franco-Cisternaa, M., Camacho-Montesc, H., & Fuentesb, M. E., (2011). Computer simulation of elastic constants of hydroxyapatite and fluorapatite. *Journal of the Mechanical Behavior of Biomedical Materials, 4*, 1011–1020.

9. Fung, Y. C., (1965). *Foundations of Solid Mechanics.* Englewood Cliffs, N.J. Prentice-Hall.

10. Hallmann, L., Mehl, A., Ulmer, P., Reusser, E., Stadler, J., & Zenobi, R., (2012). The influence of grain size on low-temperature degradation of dental zirconia. *J. Biomed. Mater. Res. B Appl. Biomat., 100*, 447–456.

11. Muhammad, I. D., et al., (2014). Modeling the elastic constants of cubic zirconia using molecular dynamics simulations. *Advanced Materials Research, 845*, 387–391.

12. Inokoshi, M., Zhang, F., De Munck, J., Minakuchi, S., Naert, I., & Vleugels, J., (2014). Influence of sintering conditions on low-temperature degradation of dental zirconia. *Dent. Mater., 30*, 669–678.

13. Kraus, H., (1980). *Creep Analysis.* Wiley, New York, USA.

14. Lubhan, D., & Felger, R. P., (1961). *Plasticity and Creep of Metals.* Wiley, New York, USA.

15. Lughi, V., & Sergo, V., (2010). Low temperature degradation-aging- of zirconia: A critical review of the relevant aspects in dentistry. *Dent. Mater., 26*, 807–820.

16. Miller, G. K., (1995). Stresses in a spherical pressure vessel undergoing creep and dimensional changes. *International Journal of Solids and Structures, 32*, 2077–2093.

17. Parkus, H., (1976). *Thermo-Elasticity.* Springer-Verlag, Wien, New York.

18. Craig, R. G., Peyton, F. A., & Johnson, D. W., (1961). Compressive properties of enamel, dental cements, and gold. *J. Dent. Res., 40*, 936–945.

19. Rimondini, L., Cerroni, L., Carrasi, A., & Torricelli, P., (2002). Bacterial colonization of zirconia ceramic surfaces: An *in vitro* and *in vivo* study. *Int. J. Oral Maxillofac Implants, 17*, 793–798.

20. Scarano, A., Piatelli, M., Caputi, S., Favero, G. A., & Piattelli, A., (2004). Bacterial adhesion on commercially pure titanium and zirconium oxide disks: An *in vivo* human study. *J. Periodontol., 75*, 292–296.

21. Seth, B. R., (1962). Transition theory of elastic-plastic deformation, creep, and relaxation. *Nature, 195*, 896–897.

22. Seth, B. R., (1966). Measure concept in mechanics. *International Journal of Non-Linear Mechanics, 1*(1), 35–40.

23. Seth, B. R., (1963). Elastic-plastic transition in shells and tubes under pressure. *Z. Angew. Math. Mech., 43*, 345–351.

24. Shahi, S., & Singh, S. B., (2020). Elastic plastic transitional stress analysis of human tooth enamel and dentine under external pressure using Seth's transition theory, materials physics, and chemistry. *Applied Mathematics and Chemo-Mechanical Analysis.* AAP, Ch: 3 (in press).

25. Shahi, S., Singh, S. B., & Thakur, P., (2019). Modeling creep parameter in rotating discs with rigid shaft exhibiting transversely isotropic and isotropic material behavior. *Journal of Emerging Technologies and Innovative Research, 6*(1), 387–395.

26. Sokolinokoff, I. S., (1956). *Mathematical Theory of Elasticity* (2nd edn.), McGraw-Hill, New York.

27. Stawarczyk, B., Ozcan, M., Hallmann, L., Ender, A., Mehl, A., & Hämmerlet, C. H., (2013). The effect of zirconia sintering temperature on flexural strength, grain size, and contrast ratio. *Clin. Oral. Investig., 17*, 269–274.

28. Thakur, P., Shahi, S., Gupta, N., & Singh, S. B., (2017). Effect of mechanical load and thickness profile on creep in a rotating disc by using Seth's transition theory. *AIP Conf. Proc., Amer. Inst. of Physics, USA, 1859*(1), 020024. doi.org/10.1063/1.4990177.

29. Timoshenko, S. P., & Woinowsky-Krieger, S., (1959). *Theory of Plates and Shells* (2nd edn.). New York: McGraw-Hill.

30. You, L. H., Zhang, J. J., & You, X. Y., (2005). Elastic analysis of internally pressurized thick-walled spherical pressure vessels of functionally graded materials. *International Journal of Pressure Vessels and Piping, 82*, 347–354.

CHAPTER 8

Recent Developments in the Theory of Nonlocality in Elastic and Thermoelastic Mediums

SUKHVEER SINGH,[1] PARVEEN LATA,[2] and SATYA BIR SINGH[3]

[1]*Assistant Professor, Punjabi University APS Neighborhood Campus, Dehla Seehan, Sangrur, Punjab, India, E-mail: Sukhveer_17@pbi.ac.in*

[2]*Associate Professor, Department of Basic and Applied Science, Punjabi University Patiala, Punjab, India*

[3]*Professor, Department of Mathematics, Punjabi University, Patiala, Punjab, India*

ABSTRACT

In the present chapter, the recent developments in the concept of nonlocality have been discussed. The nonlocal theory of continuum mechanics considers that the various physical quantities defined at a point are not just a function of the values of independent constitutive variables at that point only but a function of their values over the whole body. Some nonclassical thermoelasticity theories based on Eringen's Nonlocal theory of elasticity have been developed in the last two decades. All such theories have been very helpful in the development of the subject of thermoelasticity. A review work of all those theories and their influences on the further developments in thermoelasticity is being discussed here.

8.1 INTRODUCTION

The theory of elasticity is concerned with the study of elastic properties of a material having the property that once the deformation forces are removed, the material recovers back its original shape and size. But if the change of elastic properties of a material is due to the change in temperature, then the branch of applied mechanics dealing with this phenomenon is known as thermoelasticity. The deformations arising due to both mechanical and thermal causes were the reasons for the development of the subject of thermoelasticity. As the flow of heat through a body, causes heat conduction, stress, and strain in the body so thermoelasticity is a multi-field discipline governed by the interaction of a temperature deformation field.

If any thermodynamical changes occur in a homogeneous body in equilibrium due to the action of external loads, forces within the body, and non-uniform heating; then all such changes are concerned with the theory of Thermoelasticity. Due to such type of changes, the body undergoes deformation, thus causing stresses in the body and a rise in the body temperature. In the classical theory of thermoelasticity, the change of temperature being very small has not much effect and so the corresponding terms can be neglected. However, this is not true in such cases if the temperature undergoes a large and sudden change where the inertia term has to be considered in the equations of motion.

The nonlocal theory of thermoelasticity considers that the various physical quantities defined at a point are not just a function of the values of independent constitutive variables at that point only but a function of their values over the whole body. Thus, the nonlocal continuum field theories are concerned with the physics of material bodies in which the stress at any reference point within a continuous body not only depends on the strain at that point, but is having a significant influence due to the strains at all other points. So the nonlocal stress forces can be termed as remote action forces. Nonlocality is an essential characteristic in solid-state physics as nonlocal attractions of atoms are common. Nonlocal continuum theories can describe the material properties from microscopic scales to the size of the lattice parameter. So the nonlocal theory can satisfactorily explain some phenomena related to atomic scales.

The nonlocal theory in a way is just a generalization of the classical field theory. As: (i) the energy balance law is valid for the whole body, and (ii) the state of the body at a material point is considered to be attracted by all points of the body. This just means that a complete knowledge of the independent variables at all the points of the body is required to describe the state of the body at each point. Nonlocal effects are dominant in nature. So only, a nonlocal theory can provide the correct results where the classical theory fails. If the effects of strains at points other than the reference point are neglected, classical theory is recovered.

8.2 HISTORY AND RECENT ADVANCES IN THE CONCEPT OF NONLOCALITY

The theory of coupling of thermal and the strain fields gave rise to Coupled Theory of Thermoelasticity. Duhamel [1] derived the equations for the coupling between the strain and the temperature fields in an elastic medium, but the heat equation derived by him being parabolic gave an infinite speed of propagation for temperature. Biot [2] gave a satisfactory derivation of the heat conduction equation. Lord and Shulman [3] formulated a generalized dynamical theory of thermo-elasticity. According to this theory, the prediction is that the temperature travels with a finite speed due to the heat equation being hyperbolic.

The theory of nonlocal elasticity was developed by the contribution of many researchers. This theory is based on nonlocal effects i.e., the various physical quantities at a point are not a function of that point only but depend on their values over the whole body. Kroner [4] developed a continuum theory for long-range cohesive forces in elastic materials. He explained how the range effects can be important in materials having Vander Waals interactions as local theory gave a zero force. Edelen and Law [5] discussed a theory of nonlocal interactions and agreed to the concept of nonlocality as suggested by Kroner. Edelen et al. [6] discussed the consequences of the global postulate of energy balance and obtained the constitutive equations for the nonlinear theory. They called this nonlinear theory of nonlocal elasticity as protoelasticity.

Eringen and Edelen [7] developed the nonlocal elasticity theory via using the global balance laws and the second law of thermodynamics. This theory contains information about long-range forces of atoms as according to this theory the stress field at a particular point is affected due to the strain at all the other points of the body also. These are characterized by the presence of nonlocality residuals of fields (like body forces, internal energy, entropy, etc.). Balta and Suhubi [8] developed a new theory of nonlocal generalized thermoelasticity within the framework of the nonlocal continuum mechanics. The constitutive relations were obtained through the systematic use of the nonlocal version of the generalized thermodynamics and investigated thermal waves in rigid conductors.

Altan [9] studied the uniqueness in the linear theory of nonlocal elasticity. He concluded that under certain conditions the solutions of an initial-boundary value problem are unique. Wang and Dhaliwal [10] established a reciprocity relation and addressed some other issues addressing the nonlocal theory of micropolar elasticity to extend the concept of nonlocality further to other fields. Dhaliwal and Wang [11] provided more results on nonlocal thermoelasticity. Artan [12] compared the results of local and nonlocal elasticity theories by differentiating between the stress distributions of the local and the nonlocal theories to prove the superiority of the nonlocal theory.

The implications of the Eringen model of nonlocal elasticity theory were studied by Polizzotto [13]. He assumed an attenuation function and used it to further refine the theory. The attenuation function was supposed to be dependent on the geodetical distance replacing the Euclidean distance between material particles. He established the conditions for the uniqueness of the solution in the case of the nonlocal boundary value problem for infinitesimal displacements.

Eringen [14] presented a unified approach to the development of the basic field equations for nonlocal continuum field theories. He developed this theory for prevalent nonlocal intermolecular attractions in material bodies. In classical field theories, there exist a large number of problems falling outside their domain of applications. Some of the major deviations from the normal that confront with classical treatment are stress fields at the dislocation core, fracture of solids, short-wavelength behavior of elastic waves, etc.

Sharma and Ganti [15] described the size-dependent elastic stress state of inclusions in nonlocal media. Paola et al. [16] represented the mechanical based approach to three-dimensional nonlocal elasticity theories and proved the size dependence of the results. The model as proposed by them modeled the long-range forces as the central body forces and so for any element, the equilibrium is attained not just by contact forces at adjacent elements but also by long-range forces due to non-adjacent elements.

Ke and Wang [17] investigated the thermoelectric-mechanical vibration of the piezoelectric nanobeams-based on the nonlocal theory and Timoshenko beam theory. They discussed the influences of the nonlocal parameter, temperature change, external electric voltage, and axial force on the thermoelectric-mechanical vibration characteristics of the piezoelectric nanobeams in detail and found that the nonlocal effect is significant for the natural frequencies of the nanobeams. Khurana and Tomar [18] studied the propagation of plane longitudinal waves through an isotropic nonlocal micropolar elastic medium and showed that four dispersive waves and two sets of coupled transverse waves may propagate. As in Figure 8.1, they explained the variation of modulus of reflection coefficient with angle of incidence using local and nonlocal parameters. The solid curve denoted nonlocal micropolar solid and dotted denoted local micropolar solid.

Zenkour et al. [19] developed a generalized thermoelasticity theory with non-local deformation effects and dual-phase-lag or time-delay thermal effects to study the vibration of a nanobeam subjected to ramp-type heating. They analyzed nonlocal effects as in Figure 8.2; the transverse deflection distributions in the axial direction with different nonlocal thermoelastic parameters were studied. The dotted line indicates the value at the points where nonlocal parameter is taken to be zero, while the other two lines indicate the values with a nonlocal parameter having nonzero values, i.e., 1 and 3.

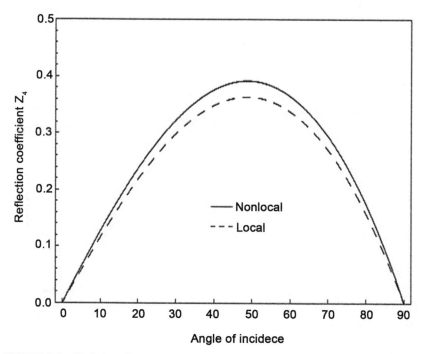

FIGURE 8.1 Variation of modulus of reflection coefficient with angle of incidence using local and nonlocal parameters.

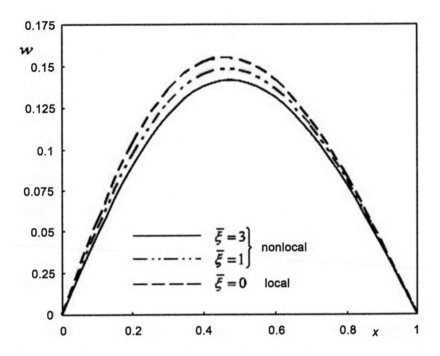

FIGURE 8.2 Transverse deflection distributions in the axial direction with different nonlocal thermoelastic parameters.

Salehipour et al. [20] gave a modified nonlocal elasticity theory stating that the strain tensor at all the neighboring points contributes to the imaginary nonlocal strain tensor at a particular point. The need for this theory arose as in the case of functionally graded materials (FGMs), the nonlocal theory of Eringen was not found to be worthy enough. They used a nonlocal strain tensor very similar to the stress tensor as used in nonlocal theory given by Eringen. This assumed strain tensor was used for obtaining the nonlocal stress tensor. Then, Vasiliev and Lurie [21] developed a new nonlocal generalized theory. Using a variational approach, they developed a new variant of nonlocal elasticity theory for generalized stresses by introducing high gradient equilibrium equations.

Jun Yu et al. [22] investigated the Buckling analysis of nanobeams with the aids of the size-dependent model, i.e., nonlocal thermoelasticity, and size effect of heat conduction. They adopted the Euler-Bernoulli beam and reformulated it within the nonlocal theory. Hayati et al. [23] analyzed the static and dynamic behavior of a curved single-walled carbon nanotube which is under twist-bending couple based on nonlocal theory. They used the nonlocal theory to model the mechanical behavior of the structure on a small scale. Togun [24] studied nonlinear vibrations of a Euler-Bernoulli nanobeam resting on an elastic foundation using nonlocal elasticity theory. Hamilton's principle was employed to derive the governing equations and boundary conditions. The nonlinear equation of motion was obtained by including stretching of the neutral axis that introduces cubic nonlinearity into the equations.

Khurana and Tomar [25] studied the propagation of Rayleigh surface waves and explored the conditions for their existence. Singh et al. [26] studied the propagation of plane harmonic waves and derived the governing relations in nonlocal elastic solid with voids. Hosseini [27] developed a meshless method based on the generalized finite difference method for coupled thermoelasticity analysis. The Green-Naghditheory of the generalized coupled thermoelasticity and nonlocal Rayleigh beam theory were utilized for dynamic analysis of a micro/nanobeam resonator subjected to thermal shock loading. Zenkour [28] used a model of nonlocal thermoelasticity theory of Green and Naghdi without energy dissipation to consider the vibration behavior of a nano-machined resonator. He used nonlocality to bring in an internal length scale in the formulation and, thus, allowing for the interpretation of size effects. The governing frequency equation was given for nanobeams subjected to different boundary conditions.

Kaur et al. [29] derived dispersion relation and investigated the propagation of Rayleigh type surface wave in a nonlocal elastic solid. They studied the nonlocal effects in detail and their effects on the propagation of Rayleigh waves and continued the good work done by Khurana et al., Tomar et al., and Singh. A part of their research work can be elaborated as in Figure 8.3, where the variation of tilt with the nonlocality parameter is studied.

Bachher and Sarkar [30] postulated a new nonlocal theory of thermoelasticity, which is based on Eringen's nonlocal elasticity theory for thermoelastic materials with voids. A material needed to be classified according to its fractional and elastic nonlocality parameter by this theory. They studied the effects of nonlocality on various parameters such as temperature, displacement, stress, and change in volume fraction. For illustration, as in Figure 8.3, they studied the effect of the nonlocality parameter on dimensionless stress. It is clear from Figure 8.4, that there is a big effect of nonlocality on stress. Similarly, he proved identical results for other parameters too.

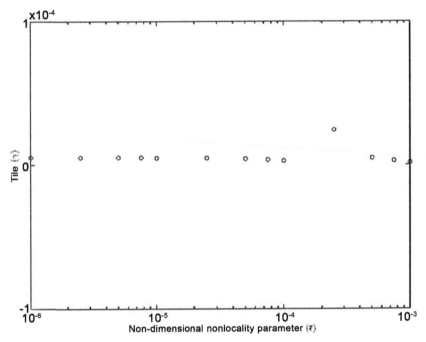

FIGURE 8.3   Variation of tilt with nonlocality parameter.

Lata [31] studied the reflection and refraction of plane waves in a layered medium of two semi-infinite nonlocal solids with the anisotropic thermoelastic medium as intermediate. She too depicted the nonlocal parameter effects graphically and proved that nonlocality is a significant parameter, which needs to be considered to get more accurate results.

8.3 CONCLUSION

The theory of elasticity and thermoelasticity are well established already. These are of wide interest in the fields where engineers, mathematicians,

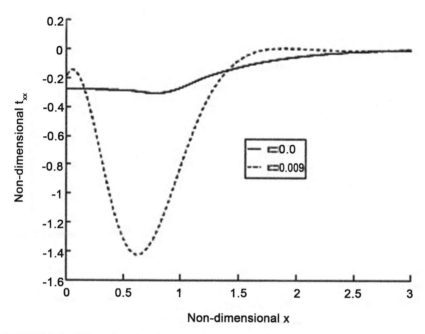

FIGURE 8.4 Effect of nonlocality parameter on dimensionless stress.

and physicists work together as these have a wider deal of applications in aeronautical engineering, nuclear physics, shipbuilding, in technologies of space vehicles and missiles, etc. As the use of the concept of nonlocality provides more refined and corrected results so it is causing more and more benefits to the subject of elasticity and thermoelasticity and thus helping in their overall development. Also, some of the well-established theories have been already refined using nonlocality and many more are to come yet. From the above, we can conclude that in the last two decades there have been great advances in the concept of nonlocality in elasticity and thermoelasticity. Many of the authors have already provided some astonishing results in the subject of thermoelasticity due to the use of nonlocality. But still, there is a lot to be done in the field of the nonlocal theory of thermoelasticity. This review chapter is a little effort to boost the researchers so that much more can be achieved in this field.

KEYWORDS

- elasticity
- Eringen model of nonlocal theories
- micropolar elastic medium
- nanobeams
- nonlocality
- thermoelasticity

REFERENCES

1. Duhamel, J. M. C., (1838). *Memoires Par Diver's Savants* (Vol. 5, pp. 440–498). Paris.
2. Biot, M. A., (1956). Thermoelasticity and irreversible thermodynamics. *Journal of Applied Physics, 27*(3), 240–253.
3. Lord, H., & Shulman, Y., (1967). A generalized dynamical theory of thermoelasticity. *Journal of the Mechanics and Physics of Solids, 15*, 299–309.
4. Kroner, E., (1967). Elasticity theory of materials with long-range cohesive forces. *International Journal of Solids and Structures, 3*, 731–742.
5. Edelen, D. G. B., & Laws, N., (1971). On the thermodynamics of systems with nonlocality. *Archive for Rational Mechanics and Analysis, 43*, 24–35.
6. Edelen, D. G. B., Green, A. E., & Laws, N., (1971). *Nonlocal Continuum Mechanics: Archive for Rational Mechanics and Analysis, 43*, 36–44.
7. Eringen, A. C., & Edelen, D. G. B., (1972). On nonlocal elasticity. *International Journal of Engineering Science, 10*, 233–248.
8. Balta, F., & Suhubi, E. S., (1977). Theory of nonlocal generalized thermoelasticity. *International Journal of Engineering and Science, 15*, 579–588.
9. Altan, S. B., (1989). Uniqueness of initial-boundary value problems in nonlocal elasticity. *International Journal of Solids and Structures, 25*(11), 1271–1278.
10. Wang, J., & Dhaliwal, R. S., (1993). On some theorems in the nonlocal theory of micro polar elasticity. *International Journal of Solids and Structures, 30*(10), 1331–1338.
11. Dhaliwal, R. S., & Wang, J., (1994). Some theorems in generalized nonlocal thermoelasticity. *International Journal of Engineering Science, 32*(3), 473–479.
12. Artan, R., (1996). Nonlocal elastic half plane loaded by a concentrated force. *International Journal of Engineering Science, 34*(8), 943–950.
13. Polizzotto, C., (2001). Nonlocal elasticity and related variational principles. *International Journal of Solids and Structures, 38*, 7359–7380.
14. Eringen, A. C., (2002). *Nonlocal Continuum Field Theories*. Springer, New York.
15. Sharma, P., & Ganti, S., (2003). The size-dependent elastic state of inclusions in non-local elastic solids. *Philosophical Magazine Letters, 83*(12), 745–754.

16. Paola, M., Failla, G., & Zingales, M., (2010). The mechanically based approach to 3D non-local linear elasticity theory: Long-range central interactions. *International Journal of Solids and Structures, 47*, 2347–2358.

17. Ke, L. L., & Wang, Y. S., (2012). Thermoelectric-mechanical vibration of piezoelectric nanobeams based on the nonlocal theory. *Smart Materials and Structures.* doi: 10.1088/0964–1726/21/2/025018.

18. Khurana, A., & Tomar, S. K., (2013). Reflection of plane longitudinal waves from the stress-free boundary of a nonlocal, microploar solid half-space. *Journal of Mechanics of Materials and Structures, 8*(1), 95–107.

19. Zenkour, A. M., Abouelregal, A. E., Alnefaie, K. A., Abu-Hamdeh, N. H., & Aifantis, E. C., (2014). A refined nonlocal thermo elasticity theory for the vibration of nano beams induced by ramp-type heating. *Applied Mathematics and Computation, 248*, 169–183.

20. Salehipour, H., Shahidi, A. R., & Nahvi, H., (2015). Modified nonlocal elasticity theory for functionally graded materials. *International Journal of Engineering Science, 90*, 44–57.

21. Vasiliev, V. V., & Lurie, S. A., (2016). On correct nonlocal generalized theories of elasticity. *Physical Mesomechanics, 19*(3), 47–59.

22. Jun, Y. Y., Xue, Z. N., Li, C. L., & Tian, X. G., (2016). Buckling of nanobeams under nonuniform temperature based on nonlocal thermoelasticity. *Composite Structures, 146*, 108–113.

23. Hayati, H., Hosseini, S. A., & Rahmani, O., (2016). Coupled twist-bending static and dynamic behavior of a curved single walled carbon nanotube based on nonlocal theory. *Micro-System Technologies.* doi: 10.1007/s00542–016–2933–0.

24. Togun, N., (2016). Nonlocal beam theory for nonlinear vibrations of a nano beam resting on elastic foundation. *Boundary Value Problems.* doi: 10.1186/s13661–016–056–3.

25. Khurana, A., & Tomar, S. K., (2017). Rayleigh type waves in nonlocal microploar solid half-space. *Ultrasonic's, 73*, 162–168.

26. Singh, D., Kaur, G., & Tomar, S. K., (2017). Waves in nonlocal elastic solid with voids. *Journal of Elasticity.* doi: 10.1007/s10659–016–9618-x.

27. Hosseini, S. M., (2017). Shock-induced nonlocal coupled thermoelasticity analysis (with energy dissipation) in a MEMS/NEMS beam resonator based on green-Naghdi theory: A mesh less implementation considering small-scale effects. *Journal of Thermal Stresses, 40*(9), 1134–1151.

28. Zenkour, A. M., (2017). Nonlocal thermo elasticity theory without energy dissipation for nano machined beam resonators subjected to various boundary conditions. *Micro-System Technologies.* doi: 10.1007/s00542–015–2703–4.

29. Kaur, G., Singh, D., & Tomar, S. K., (2018). Rayleigh type wave in nonlocal elastic solid with voids. *European Journal of Mechanics.* doi: 10.1016/j.euromechsol.2018.03.015.

30. Bachher, M., & Sarkar, N., (2018). Nonlocal theory of thermoelastic materials with voids and fractional derivative heat transfer. *Waves in Random and Complex Media.* doi: 10.1080/17455030.2018.1457230.

31. Lata, P., (2018). Reflection and refraction of plane waves in layered nonlocal elastic and anisotropic thermo elastic medium. *Structural Engineering and Mechanics, 66*(1), pp. 113–124.

32. Sharma, N., Kumar, R., & Lata, P., (2015). Disturbance due to inclined load in the transversely isotropic thermo elastic medium with two temperatures and without energy dissipation. *Material Physics and Mechanics, 22*, 107–117.

CHAPTER 9

Investigation of Creep Performance of an Isotropic Composite Disc Under Thermal Gradients

VANDANA GUPTA[1] and SATYA BIR SINGH[2]

[1]*Department of Mathematics, Dashmesh Khalsa College, Zirakpur, Mohali, Punjab, India, E-mail: vaggarwal2584@gmail.com*

[2]*Department of Mathematics, Punjabi University, Patiala, Punjab, India*

9.1 INTRODUCTION

A rotating disc is a common component in turbines, gas turbine rotors, compressors, turbo generators, flywheel, aero-engines, rotors, automobiles, computer's disc drive, and other dynamic applications are usually operated at relatively higher angular speed and subjected to high temperature. Materials of discs are required to sustain steady loads for a long period of time under different temperature conditions. In such conditions, the material may continue to creep until its usefulness is seriously impaired. As a result, a number of researchers have studied the creep behavior in composite materials with the property of superior heat resistance. The effect of anisotropy on the stress and strain rates have been studied and concluded that the anisotropy of the material has a significant effect on the creep of a rotating disc [1]. The steady-state creep response of an isotropic FGM disc with constant thickness by using Sherby's constitutive model have been analyzed and the results obtained for nonlinear variation of particle distribution along the radial distance of the disc are compared with that of discs containing the same amount of particle distributed uniformly or linearly along with the radial distance [17]. Hasan Callioglu et al. [18] studied that stress analysis on functionally graded rotating annular discs subjected to temperature distributions parabolically decreasing with radius. He concluded that with the increase in temperature, the tangential stress component gets decreased at the inner surface but increased at the outer surface whereas the radial stress component gets reduced gradually for all the temperature distributions. The problem of rotating disc with variable thickness, thermal effect, heat generation effect, and pressure by using Seth's transition theory has been studied [2, 19–21]. Gupta et al. [3] discuss the variation of Poisson ratios and thermal creep stresses and strain rates in an isotropic disc. Kaur et al. [4] study steady thermal stresses in a thin rotating disc of infinitesimal deformation with edge loading. An attempt has been made to investigate steady-state creep behavior of thermally graded isotropic discs rotating at elevated temperatures. The results are compared with the disc having a uniform temperature profile from inner to the outer radius and are displayed graphically in a designer-friendly format for the said temperature profiles. It is observed that there is a need to extend the domain of thermal gradation for designing rotating discs [22]. A theoretical solution for time-dependent thermo-elastic creep analysis of a functionally graded thick-walled cylinder based on the first-order shear deformation theory is presented. The effects of the temperature gradient and FG grading index on the creep stresses of the cylinder are investigated. A numerical solution using the finite element method is also presented and good agreement was found [5]. Keeping this in mind, the study ends with an effort to determine the plastic stress and strain analyses for the particle reinforced isotropic/anisotropic disc with constant thickness in presence of thermal gradients and compare it with those isotropic/anisotropic discs with the operating under isothermal conditions. The analysis has been done by using von Mises criteria/Hill's criterion for yielding. The creep response of the disc under stresses developing due to rotation has been determined using Sherby's law. The material parameters of creep vary along the radial direction in the disc due to varying composition.

Keeping this in mind, the study ends with an effort to determine the creep behavior for the particle reinforced anisotropic disc with constant thickness in presence of thermal gradients and compare it with those anisotropic discs with the operating under isothermal conditions.

In this chapter, the steady-state creep has been analyzed in composite disc made of material 6061Al base alloy containing 20 vol% of SiC particle. The creep behavior has been described by Sherby's constitutive model. The analysis has been done for particle reinforced isotropic disc with hyperbolic thickness.

9.2 DISC GEOMETRY AND REINFORCEMENT PROFILE

The present study assumes composite disc having a hyperbolically varying thickness is assumed to be of the form:

$$h = cr^m \tag{9.1}$$

where: c and $m = -0.74$ are constants.

Since the volume of disc having hyperbolically varying thickness is assumed equal to the volume of constant thickness disc, therefore;

$$c = \frac{(m+2)(b^2 - a^2)t}{2(b^{m+2} - a^{m+2})} \tag{9.2}$$

The thickness of the disc is assumed to be h and a and b be inner and outer radii of the disc respectively. Let I and I_0 be the moment of inertia of the disc at inner radius a and outer radius r and b, respectively. A and A_0 be the area of cross-section of the disc at inner radius a and outer radius r and b, respectively.

Here,

$$A = c\left(\frac{r^{m+1} - a^{m+1}}{m+1}\right) \tag{9.3}$$

$$A_0 = c\left(\frac{b^{m+1} - a^{m+1}}{m+1}\right) \tag{9.4}$$

$$I = c\left(\frac{r^{m+3} - a^{m+3}}{m+3}\right) \tag{9.5}$$

$$I_0 = c\left(\frac{b^{m+3} - a^{m+3}}{m+3}\right) \tag{9.6}$$

The temperature gradient originating due to the braking action of the discs has been obtained by Finite Element Analysis. The temperature $T(r)$, obtained at any radius r is presented below in the form of regression equation as:

$$T(r) = a_0 + a_1 r + a_2 r^2 + a_3 r^3 + a_4 r^4 + a_5 r^5 \qquad (9.7)$$

where the coefficients a_0, a_1, a_2, a_3, a_4 and a_5 for different disc are given below:

$$a_0 = 602.75,\ a_1 = 1.4488$$

$$a_2 = -0.0208,\ a_3 = 3.27 \times 10^{-4}$$

$$a_4 = 1.96 \times 10^{-6},\ a_5 = 4.43 \times 10^{-9}$$

9.3 ASSUMPTIONS IN VARIABLE THICKNESS DISC

Consider an aluminum silicon-carbide particulate composite disc of constant thickness h having an inner radius, a and outer radius, b rotating with angular velocity, ω(radian/sec). From symmetry considerations, principal stresses are in the radial, tangential, and axial directions. For the purpose of analysis, the following assumptions are made:

1. Stresses at a radius of the disc remain constant with time, i.e., steady-state condition of stress is assumed.
2. Elastic deformations are small for the disc and can be neglected as compared to the creep deformations.
3. Biaxial state of stress ($\sigma_z = 0$) exists at any point in the disc.
4. Frictional shear stress-induced due to braking action is estimated to be 10^5 MPa, which is very small compared to creep stresses and therefore, can be neglected.
5. The composite shows a steady-state creep behavior which may be described by following Sherby's law [6],

$$\dot{\bar{\varepsilon}} = \left[M(r)\left(\bar{\sigma} - \sigma_0(r)\right)\right]^n \qquad (9.8)$$

where, $M(r) = \dfrac{1}{E}\left(\dfrac{AD_\lambda \lambda^3}{|b_r|^5}\right)^{1/n}$ is the creep parameter and $\dot{\bar{\varepsilon}}, \bar{\sigma}, n, \sigma_0(r)$,

$A, D_\lambda, \lambda, b_r, E$ are the effective strain rate, effective stress, the stress exponent, threshold stress, a constant, lattice diffusivity, the subgrain size, the magnitude of burgers vector, Young's modulus.

The values of material parameters $M(r)$ and $\sigma_0(r)$ in terms of $P, T(r)$ and V have been obtained from the creep results by using the experimental results reported by Pandey et al. [7] for Al-SiCp composite under uniaxial loading using the following regression equations:

$$M(r) = e^{-35} \, P^{\,0.2077} \, T(r)^{\,4.98} \, V^{\,-0.622} \qquad (9.9)$$

$$\sigma_0(r) = -0.03507P + 0.01057T(r) + 1.00536 - 2.11916 \qquad (9.10)$$

In a FGM disc, with the creep parameters $M(r)$ and $\sigma_0(r)$ will vary radially due to variations in temperature $T(r)$. In the present study, the particle size (P) and the particle content (V) are taken as 1.7 μm and 20% over the entire disc. Thus, for a given FGM disc under known temperature both the creep parameters are functions of radial distance and their values $M(r)$ and $\sigma_0(r)$ at any radius (r), could be determined by substituting the values of particle size, particle content, and temperature distributions into Eqs. (9.9) and (9.10), respectively.

9.4 MATHEMATICAL FORMULATION

The generalized constitutive equations for creep in an anisotropic composite disc under multiaxial stress takes the following form:

$$\dot{\varepsilon}_r = \frac{\dot{\bar{\varepsilon}}}{2\bar{\sigma}}\left[2\sigma_r - \left(\sigma_\theta + \sigma_z\right)\right] \qquad (9.11)$$

$$\dot{\varepsilon}_\theta = \frac{\dot{\bar{\varepsilon}}}{2\bar{\sigma}}\left[2\sigma_\theta - \left(\sigma_z + \sigma_r\right)\right] \qquad (9.12)$$

$$\dot{\varepsilon}_z = \frac{\dot{\bar{\varepsilon}}}{2\bar{\sigma}}\left[2\sigma_z - \left(\sigma_r + \sigma_\theta\right)\right] \qquad (9.13)$$

where $\dot{\varepsilon}_r, \dot{\varepsilon}_\theta, \dot{\varepsilon}_z$ and $\sigma_r, \sigma_\theta, \sigma_z$ are the strain rates and the stresses respectively in the direction r, θ and z. $\dot{\bar{\varepsilon}}$ be the effective strain rate and $\bar{\sigma}$ be the effective stress. For biaxial state of stress $(\sigma_r, \sigma_\theta)$ the effective stress is:

$$\bar{\sigma} = \left\{ \frac{1}{2} \left\{ \sigma_\theta^2 + \sigma_r^2 + \left(\sigma_r - \sigma_\theta \right)^2 \right\} \right\}^{1/2} \tag{9.14}$$

Using Eqs. (9.8) and (9.14), Eq. (9.11) can be rewritten as:

$$\dot{\varepsilon}_r = \frac{d\dot{u}_r}{dr} = \frac{\left[2x(r) - 1 \right] \left[M\left(\bar{\sigma} - \sigma_0 \right) \right]^8}{2 \left[x^2(r) - x(r) + 1 \right]^{1/2}} \tag{9.15}$$

Similarly from Eq. (9.12):

$$\dot{\varepsilon}_\theta = \frac{\dot{u}_r}{r} = \frac{\left[2 - x(r) \right] \left[M\left(\bar{\sigma} - \sigma_0 \right) \right]^8}{2 \left[x^2(r) - x(r) + 1 \right]^{1/2}} \tag{9.16}$$

$$\dot{\varepsilon}_z = -\left(\dot{\varepsilon}_r + \dot{\varepsilon}_\theta \right) \tag{9.17}$$

where, $x(r) = \dfrac{\sigma_r}{\sigma_\theta}$ is the ratio of radial and tangential stresses at any radius r.

Dividing Eq. (9.15) by Eq. (9.16), we get:

$$\phi(r) = \frac{2x(r) - 1}{2 - x(r)} \tag{9.18}$$

The equation of equilibrium for a rotating disc with varying thickness can be written as:

$$\frac{d}{dr}(r\sigma_r) - \sigma_\theta + \frac{\rho(r)\omega^2 r^2}{g} = 0 \tag{9.19}$$

where $\rho(r)$ is the density of FGM disc.

Boundary conditions are:

$$\sigma_r(a) = 0 = \sigma_r(b) \tag{9.20}$$

We get the tangential stress σ_θ from Eq. (9.19) by using Eq. (9.15) and Eq. (9.16):

$$\sigma_\theta = \frac{\psi_1(r)\left[A_0\,\sigma_{\theta avg} - \int_a^b \dot{E}_2(r).dr\right]}{M(r)\int_a^b \dfrac{\psi_1(r)}{M(r)}dr} + \psi_2(r) \tag{9.21}$$

where,

$$\psi_1(r) = \frac{\psi(r)}{\left[x^2(r) - x(r) + 1\right]^{1/2}} \tag{9.22}$$

$$\psi_2(r) = \frac{\sigma_0}{\left[x^2(r) - x(r) + 1\right]^{1/2}} \tag{9.23}$$

and

$$\psi(r) = \left\{\frac{2}{r} \cdot \frac{\left[x^2(r) - x(r) + 1\right]^{1/2}}{\left[2 - x(r)\right]} \cdot \int_a^r \frac{Æ(r)dr}{r}\right\}^{1/8} \tag{9.24}$$

The average tangential stress may be defined as:

$$\sigma_{\theta avg} = \frac{1}{A_0}\rho\omega^2 I_0 \tag{9.25}$$

Now $\sigma_r(r)$ can be obtained by integrating Eq. (9.18) within limits a to b:

$$\sigma_r(r) = \frac{1}{r}\left[\int_a^r \sigma_\theta \, dr - \frac{\omega^2 \rho\left(r^3 - a^3\right)}{3}\right] \qquad (9.26)$$

Thus, the tangential stress σ_θ and radial stress σ_r are determined by Eqs. (9.21) and (9.26) respectively, for anisotropic disc with constant thickness.

Then strain rates $\dot{\varepsilon}_r, \dot{\varepsilon}_\theta$ and $\dot{\varepsilon}_z$ calculated from equations (9.15), (9.16) and (9.17).

9.5 NUMERICAL COMPUTATION

The stress distribution is evaluated from the above analysis by the iterative numerical scheme of computation. For rapid convergence, 75% of the value of $\sigma_\theta(r)$ obtained in the current iteration has been mixed with 25% of the value of $\sigma_\theta(r)$ obtained in the last iteration for the use in the next iteration.

9.6 RESULTS AND DISCUSSION

The stress exponent and density of disc material have been taken as $n = 8$ and $\rho = 2812.4 kg/m^3$, respectively. A computer program based on the mathematical formulation has been developed to obtain the steady-state creep response of the rotating isotropic discs operating under the thermal gradient. The tangential stress operating under a thermal gradient is a little higher near the inner radius and slightly lowers near the outer radius as compared to the disc without thermal gradients as shown in Figure 9.1.

The radial stress developing due to rotation in the isotropic disc operating under a thermal gradient is higher over the entire radius as compared to the isotropic disc without thermal gradients as shown in Figure 9.2. Although, the variation in the magnitude of tangential as well as radial stress distribution is very small in the isotropic discs due to the presence of thermal gradients. In Figure 9.3, the tangential strain rate decreases significantly over the entire radius in the isotropic discs operating under the thermal gradients compared to the discs without thermal gradients. Although, the tangential stress in disc under the thermal gradients is higher near the inner radius than the disc under thermal gradients as shown in Figure 9.4, but lower

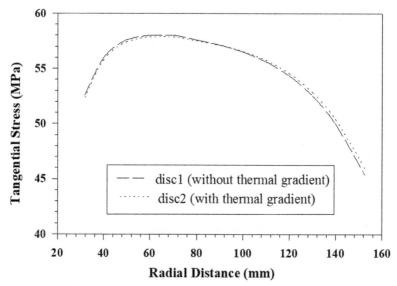

FIGURE 9.1 Variation of tangential stress along with radial distance in composite hyperbolically varying discs with/without thermal gradient.

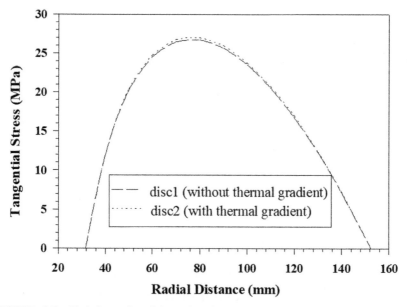

FIGURE 9.2 Variation of radial stress along with radial distance in composite hyperbolically varying discs with/without thermal gradient.

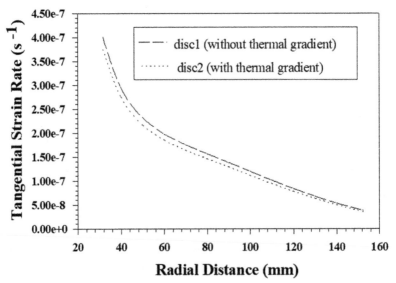

FIGURE 9.3 Variation of tangential strain rate along with radial distance in composite hyperbolically varying discs with/without thermal gradient.

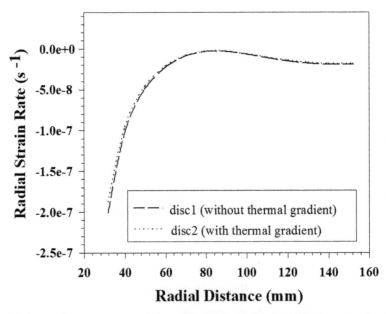

FIGURE 9.4 Variation of radial strain rate along with radial distance in composite discs with/without thermal gradient.

operating temperature near the inner radius of the disc with thermal gradients is able to reduce the creep rate overcoming the effect of higher stress. In Figure 9.4, the effect of imposing thermal gradients on the radial strain rate in the isotropic discs is similar to that observed for the tangential strain rate. The magnitude of the radial strain rate firstly increases rapidly with radial distance and then starts decreasing.

9.7 CONCLUSION

From the above discussion, it can be concluded that:

1. The thermal gradients significantly affect the strain rate distribution in an isotropic particle reinforced disc having a hyperbolic thickness, but its effect on the distribution of stresses is relatively small.
2. The thermal gradient plays a significant role in developing the creep strains, it may be taken care of while design a rotating disc.

KEYWORDS

- **anisotropy**
- **creep strains**
- **reinforcement profile**
- **rotating disc**
- **silicon carbide particles**
- **thermal gradients**

REFERENCES

1. Chamoli, N., Rattan, M., & Singh, S. B., (2010). Effect of anisotropy on the creep of a rotating disc of Al-SiCp composite. *International Journal of Contemporary Mathematical Sciences, 5*(11), 509–516.
2. Thakur, P., Singh, S. B., Singh, J., & Kumar, S., (2016a). Steady Thermal stresses in solid disk under heat generation subjected to variable density. *Kragujevac Journal of Science, 38*, 5–14.

3. Gupta, N., Singh, S. B., Thakur, P., (2016). Determine variation of Poisson ratios and thermal creep stresses and strain rates in an isotropic disc. *Kragujevac Journal of Science, 38*, 15–28.

4. Kaur, J., Thakur, P., & Singh, S. B., (2016). Steady thermal stresses in a thin rotating disc of finitesimal deformation with edge loading. *Journal of Solid Mechanics, 8*(1), 204–211.

5. Kashkoli, M. D., Tahan, K. N., & Nejad, M. Z., (2017). Time-dependent thermo mechanical creep behavior of FGM thick hollow cylindrical shells under non-uniform internal pressure. *International Journal of Applied Mechanics, 9*(6), 1–26, Article number: 1750086.

6. Sherby, O. D., Klundt, R. H., & Miller, A. K., (1977). Flow stress, sub grain size and subgrain stability at elevated temperature. *Metallurgical Transactions, 8A*, 843–850.

7. Pandey, A. B., Mishra, R. S., & Mahajan, Y. R., (1992). Steady state creep behavior of silicon carbide particulate reinforced aluminum composites. *Acta Metallurgica. Materialia., 40*(8), 2045–2052.

8. Arya, V. K., & Bhatnagar, N. S., (1979). Creep analysis of rotating orthotropic disc. *Nuclear Engineering and Design, 55*, 323–330.

9. Bhatnagar, N. S., Kulkarni, P. S., & Arya, V. K., (1986). Steady state creep of orthotropic rotating discs of variable thickness. *Nuclear Engineering and Design, 91*(2), 121–141.

10. Deepak, D., Gupta, V. K., & Dham, A. K., (2013). Investigating the effect of thickness profile of a rotating functionally graded disc on its creep behavior. *Journal of Thermoplastic Composite Materials, 26*(4), 461–475.

11. Durodola, J. F., & Attia, O., (2000). Deformation and stresses in functionally graded rotating discs. *Composites Science and Technology, 60*(7), 987–995.

12. Gupta, V. K., Singh, S. B., Chandrawat, H. N., & Ray, S., (2004). Steady state creep and material parameters in a rotating disc of Al-SiCp composite. *European Journal of Mechanics A/Solids, 23*(3), 335–344.

13. Gupta, V. K., Singh, S. B., Chandrawat, H. N., & Ray, S., (2005). Modeling of creep behavior of a rotating disc in presence of both composition and thermal gradients. *Journal of Engineering Materials and Technology, 127*(1), 97–105.

14. Mishra, R. S., & Pandey, A. B., (1990). Some observations on the high-temperature creep behavior of 6061 Al-SiC composites. *Metallurgical Transactions, 21A*(7), 2089–2091.

15. Rattan, M., Bose, T., & Chamoli, N., (2016). Effect of linear thermal gradient on steady-state creep behavior of isotropic rotating disc. *International Journal of Mechanical and Mechatronics Engineering, 11*(5), 1067–1073.

16. Wahl, A. M., Sankey, G. O., Manjoine, M. J., & Shoemaker, E., (1954). Creep tests of rotating risks at elevated temperature and comparison with theory. *Journal of Applied Mechanics, 76*, 225–235.

17. Singh, S. B., & Rattan, M. (2010). Creep Analysis of an Isotropic Rotating Al—SiCp Composite Disc taking into Account the Phase-specific Thermal Residual Stress. *Journal of Thermoplastic Composite Materials, 23*(3), 299–312.

18. Çallioğlu, H., Bektaş, N. B., & Sayer, M. (2011). Stress analysis of functionally graded rotating discs: analytical and numerical solutions. *ActaMech Sin, 27*, 950–955.

19. Thakur, P. (2013). *Thermo Creep Transition Stresses in a Thick-Walled Cylinder subjected to Internal Pressure, Structure Integrity and Life, Serbia, 12*(3), 165–173.
20. Thakur, P. (2015). *Analysis of Thermal Creep Stresses in Transversely Thick-Walled Cylinder subjected to Pressure, Structure Integrity and Life, Serbia, 15*(1), 19–26.
21. Thakur, P., & Kumar, S. (2016). *Stress Evaluation in a Transversely Isotropic Circular Disk with an Inclusion, Structural Integrity and Life, 16*(3), 155–160.
22. Tania Bose, Minto Rattan, & Neeraj Chamoli (2017). Steady State Creep of Isotropic Rotating Composite Disc Under Thermal Gradation, *International Journal of Applied Mechanics, 9*(6), 1750077.

CHAPTER 10

Application of a Fractional-Order PID Controller to an Underactuated System

TUNG-YUNG HUANG and SHIH-YING HUNG

Department of Mechanical Engineering,
Southern Taiwan University of Science and Technology, Tainan,
Taiwan, E-mail: huangt@stust.edu.tw (T. Y. Huang)

10.1 INTRODUCTION

The governing equations of dynamic systems or processes can be generalized by arbitrary or fractional-order differential equations including integer-order ones, therefore it would be suitable to apply fractional calculus to controller design. The concept of fractional calculus is first mentioned in a letter Leibniz wrote to de L'Hopital in 1695. And its applications may be dated back to the early 19th century, Abel uses fractional calculus to formulate the "tautochrone" problem and find its solution [1]. Later on, researchers such as Liouville and Heaviside also use fractional calculus to derive mathematical models for systems including potential theory, heat equation, and notch design of a dam, and solve those problems therewith. Then the related research seemed to have stalled inexplicably until the late 20th century. Researches on miscellaneous fractional-order systems, including rheology, viscoelasticity, diffusion, electromagnetic theory, electrochemistry of corrosion, and statistics have been made during the past several decades. Inevitably, researchers would apply the technique to control systems.

The well-known PID controller is typical of controllers used in the industry. It uses the linear combination of the input error value, and it's integral and derivative as the output. One advantage of PID control is that no complex algorithm or calculation is involved. In addition, there are empirical parameters tuning methods designed for it, such as the

Ziegler-Nichols method [2] which requires only the system's time response without knowing its parametric model. The simplicity and effectiveness of these tuning methods make the PID controller even more appealing. It is desired to introduce the fractional calculus technique to such a controller and find a suitable tuning rule for the adapted controller, namely the fraction order PID controller.

Most researchers propose to optimize fractional-order PID controllers' parameters by using gain crossover frequency and phase margin [3]. Yet, this method requires fairly good initial guess so as to have the parameters converge to optimum. Valério and Sá da Costa [4] propose a scheme for tuning fractional-orders PID controllers' parameters adhering to the concept of the Ziegler-Nichols method. However, it's useful only when step response with time delay in sigmoidal shape, not suitable for system response without time delay.

This research studied the feasibility of applying the fractional-order PID controller (a.k.a. $PI^{\lambda}D^{\mu}$ controller, $\lambda, \mu > 0$) to a DC motor-driven two-link rotary pendulum system, an underactuated system. Some researchers also apply the fractional-order PID controller to underactuated systems such as an overhead 2D crane and an inverted pendulum on a cart [5, 6] without any specific approach. In view of the simplicity and effectiveness of the Ziegler-Nichols method, a variant of the Ziegler-Nichols method is proposed to determine the parameters of the fractional-order PID controller in this research. Different order $PI^{\lambda}D^{\mu}$ controllers are tried for comparison in order to find a $PI^{\lambda}D^{\mu}$ controller of appropriate order for the specified system.

This chapter consists of the following sections: Section 10.1 is the introduction, Section 10.2 briefly introduces the fractional calculus, fractional-order PID controllers, and the proposed tuning method, Section 10.3 presents the modeling of a DC motor-driven two-link rotary pendulum system, Section 10.4 estimates the controller parameters by using local linearization and the Ruth-Hurwitz stability criterion, Section 10.5 gives the simulation results, and Section 10.6 is the conclusion.

10.2   FRACTIONAL-ORDER PID CONTROLLERS

10.2.1   *FRACTIONAL CALCULUS*

Fractional calculus operator is defined as follows [7]:

$$_aD_t^\alpha \begin{cases} \dfrac{d^\alpha}{dt^\alpha} & \mathrm{Re}(\alpha) > 0 \\ 1 & \mathrm{Re}(\alpha) = 0 \\ \int_a^t (d\tau)^{-a} & \mathrm{Re}(\alpha) < 0 \end{cases} \qquad (10.1)$$

where a and t are the lower limit and upper limit of the integral, respectively, and α is the fractional order, which will be taken as a real number afterwards. If $\alpha > 0$, $_aD_t^\alpha$ is a fractional-order differential, and if $\alpha < 0$, $_aD_t^\alpha$ is a fractional-order integral. Note that when the infinitesimal increment h in time equals $t-a$, both the left and right subscripts of the D operator can be omitted, and when α is an integer, the operator functions as the commonly seen integer D operator, that is, $D^k = d^k/dt^k \ (k \in N)$ as a k-th order differentiator, and $D^{-1} = \int d\tau$ for the first order integral and $D^{-k} = \int \int ... \int d\tau...d\tau d\tau \ (k \text{ times } k \in N)$ for the k-th order integral.

One most used fractional-order differential definition is the Grünwald-Letnikov definition [7], which is expressed as follows:

$$_aD_t^\alpha f(t) = \lim_{\substack{h \to 0 \\ nh=t-a}} \left(h^{-\alpha} \sum_{r=0}^{n} (-1)^r \binom{\alpha}{r} f(t-rh) \right) \qquad (10.2)$$

where

$$\binom{\alpha}{r} = \frac{\alpha(\alpha-1)\cdots(\alpha-r+1)}{r!} \qquad (10.3)$$

with

$$\binom{\alpha}{0} = 1$$

Meanwhile a famous fractional-order integral definition is the Riemann-Liouville definition [7], which is expressed as follows:

$$_aD_t^\alpha f(t) = \frac{1}{\Gamma(\alpha)} \int_a^t (t-\tau)^{\alpha-1} f(\tau) d\tau \qquad (10.4)$$

where

$$\Gamma(x) = \int_0^\infty y^{x-1} e^{-y} \, dy$$

(10.5)

is the well-known Euler's gamma function.

As engineering problems described by ordinary differential equations in integer order can be solved using the Laplace transform, so are those systems in fractional-order differo-integral equations.

According to the Riemann-Liouville definition, the Laplace transform for fractional-order differo-integrals is:

$$\mathcal{L}\{{_0D_t^\alpha f(t)}\} = s^\alpha F(s) - \sum_{j=0}^{n-1} s^j \left[{_0D_t^{\alpha-j-1} f(0)}\right]$$

(10.6)

when $(n-1) \leq \alpha < n$ If ${_0D_t^{\alpha-j-1} f(0)} = 0$, $j= 0, 1, 2, \dots, n-1$, then

$$\mathcal{L}\{{_0D_t^\alpha f(t)}\} = s^\alpha F(s)$$

Now the order of the Laplace operator s is a fraction in fractional-order differo-integral.

10.2.2 PID CONTROLLERS AND ZIEGLER-NICHOLS METHOD

10.2.2.1 PID CONTROLLERS

PID controllers have been popularly used after its debut. Its control effort $u(t)$ in time domain can be expressed as:

$$u(t) = K_P e(t) + K_I \int e(t) \, dt + K_D \frac{de(t)}{dt}$$

(10.7)

where K_P, K_I and K_D are the proportional gain, the integral gain, and the derivative gain, respectively. And the corresponding transfer function obtained by taking the Laplace transform of Eq. (10.7) is:

$$G_C(s) = \frac{U(s)}{E(s)} = K_P + \frac{K_I}{s} + K_D s$$

(10.8)

Alternatively, the parameter set is defined somehow in a different way as follows:

$$u(t) = K_P \left(e(t) + \frac{1}{T_I} \int e(t)dt + T_D \frac{de(t)}{dt} \right) \tag{10.9}$$

And the corresponding transfer function of Eq. (10.9) is:

$$G_C(s) = \frac{U(s)}{E(s)} = K_P \left(1 + \frac{1}{T_I s} + T_D s \right) \tag{10.10}$$

By comparison of these two expressions, it is easy to conclude that $K_I = K_p/T_I$, and $K_D = K_P T_D$. Furthermore, it's noteworthy that the 3 terms inside the parenthesis of Eq. (10.9) have the same dimension of $e(t)$, where the unit of T_I and T_D (both of physical quantity "time") is "second" used to cancel the unit of "dt" in the integral and derivative, respectively.

10.2.2.2 *ZIEGLER-NICHOLS METHOD*

The Ziegler-Nichols method [2] is a heuristic method for tuning PID controllers. It is meant to make the closed-loop control system's response acceptable though not optimal. The procedure of the Ziegler-Nichols method is first to set the integral gain and derivative gain to zero which results in a P controller. Then vary the proportional gain to yield a sustained periodic oscillation in the output (or close to it) by trial and error. Mark the critical gain as the ultimate gain K_u and the corresponding period as the ultimate period T_u. Then, according to the Ziegler-Nichols method, the empirical PID parameters are assigned to be $K_p = 0.6K_u$, $T_I = T_u/2$, and $T_D = T_u/8$, which is equivalent to have the alternative parameter set: $K_p = 0.6K_u$, $K_I = 1.2K_u/T_u$, and $K_D = 3K_u T_u/40$.

10.2.3 *FRACTIONAL-ORDER PID CONTROLLERS AND TUNING RULE*

10.2.3.1 *FRACTIONAL-ORDER PID CONTROLLERS*

Oustalop [8] proposes to apply fractional-order controllers to dynamic systems, and names such robust controllers as "Commande Robuste d'Ordre Non-Entier" (CRONE). He also introduces the fractional-order PID controller, or the PI$^\lambda$D$^\mu$ controller (λ and μ are non-negative fractions). Similar to the PID controllers, the control effort of the fractional-order PID controller $u(t)$ in the time domain can be expressed as:

$$u(t) = K_P\, e(t) + K_{I,\lambda}\, D^{-\lambda} e(t) + K_{D,\mu}\, D^{\mu} e(t)$$

$$= K_P\, e(t) + K_{I,\lambda} \int e(t)(dt)^{\lambda} + K_{D,\mu} \frac{d^{\mu} e(t)}{dt^{\mu}} \qquad (10.11)$$

where K_p, $K_{I,\lambda}$, and $K_{D,\mu}$ are the proportional gain, the fractional-order integral gain, and the fractional-order derivative gain, respectively.

And the corresponding transfer function takes the following form:

$$G_C(s) = K_P + \frac{K_{I,\lambda}}{s^{\lambda}} + K_{D,\mu} s^{\mu} \qquad (10.12)$$

when $\lambda = \mu = 1$ in Eq. (10.12), $G_C(s)$ is an integer PID controller; $\lambda = 1$, $\mu = 0$, $G_C(s)$ an integer PD controller; $\lambda = 0$, $\mu = 1$, $G_C(s)$ an integer PI controller; and $\lambda = \mu = 0$, $G_C(s)$ an integer P controller. Without limiting λ and μ to be 0 or 1, the fractional-order differo-integral offers more flexibility than the integer-order differo-integral in the controller design.

10.2.3.2 TUNING RULE: GENERALIZED ZIEGLER-NICHOLS METHOD

It is desired to find a general tuning rule, just like the Ziegler-Nichols method, of setting empirical values for fractional-order PID controllers' parameters. As mentioned in Section 10.2.2.1, the three terms inside the parenthesis of Eq. (10.9) are of the same dimension, i.e., the unit of T_I and T_D cancels the unit of "dt" in the integral and derivative, respectively. Keeping this in mind, now the orders of the fractional integral and derivative are λ and μ in Eq. (10.11), respectively. In order to cancel the unit of "$(dt)^{\lambda}$" in the integral and "$(dt)^{\mu}$" in the derivative, $(T_I)^{\lambda}$ and $(T_D)^{\mu}$ can be employed in the control law for the fractional integral and derivative, respectively. Therefore, the control law for the fractional PID controller can be expressed as:

$$u(t) = K_P\, e(t) + K_{I,\lambda} \int e(t)(dt)^{\lambda} + K_{D,\mu}\, \frac{d^{\mu}e(t)}{dt^{\mu}}$$

$$= K_P\left(e(t) + \frac{1}{T_I^{\lambda}} \int e(t)(dt)^{\lambda} + T_D^{\mu}\, \frac{d^{\mu}e(t)}{dt^{\mu}} \right)$$

$$= K_P\left(e(t) + \left(\frac{2}{T_u}\right)^{\lambda} \int e(t)(dt)^{\lambda} + \left(\frac{T_u}{8}\right)^{\mu}\, \frac{d^{\mu}e(t)}{dt^{\mu}} \right) \qquad (10.13)$$

which yields $K_{I,\lambda} = \dfrac{K_P}{T_I^{\lambda}} = K_P\left(\dfrac{2}{T_u}\right)^{\lambda}$ and $K_{D,\mu} = K_P T_D^{\mu} = K_P\left(\dfrac{T_u}{8}\right)^{\mu}$.

Note that $\lambda = \mu = 1$ is a special case of the fractional-order PID controller, and indeed it becomes the commonly seen PID controller with parameters set by the Ziegler-Nichols method. Hence Eq. (10.13) generalizes the integer-order PID controller where $\lambda = \mu = 1$.

10.2.4 IMPLEMENTATION OF FRACTIONAL-ORDER PID CONTROLLERS

10.2.4.1 APPROXIMATION OF FRACTIONAL-ORDER LAPLACE OPERATOR

The fractional-order differentiator s^r (r is a positive fraction) in continuous-time domain can be expressed in terms of z^{-1} in discrete-time domain by a generating function $s = w(z^{-1})$ just like the integer-order differentiator s (z is an operator in Z transform); similarly it applies to the fractional-order integrator s^{-r} (r is a positive fraction). Adopting the Tustin generating function, the fractional differero-integral would be discretized as [9]:

$$s^{\pm r} = \left(w(z^{-1})\right)^{\pm r} = \left(\frac{2}{T}\left(\frac{1-z^{-1}}{1+z^{-1}}\right)\right)^{\pm r} \qquad (10.14)$$

where $\left(\dfrac{1-z^{-1}}{1+z^{-1}}\right)^r$ can be expanded in Taylor series as follows:

$$\left(\frac{1-z^{-1}}{1+z^{-1}}\right)^r = 1 - (2r)z^{-1} + (2r^2)z^{-2} - \left(\frac{2}{3}r + \frac{4}{3}r^3\right)z^{-3}$$
$$+ \left(\frac{4}{3}r^2 + \frac{2}{3}r^4\right)z^{-4} - \left(\frac{2}{5}r + \frac{4}{3}r^3 + \frac{4}{15}r^5\right)z^{-5} + \cdots \tag{10.15}$$

Therefore,

$$s^r = \left(\frac{2}{T}\right)^r \Bigg(1 - (2r)z^{-1} + (2r^2)z^{-2} - \left(\frac{2}{3}r + \frac{4}{3}r^3\right)z^{-3}$$
$$+ \left(\frac{4}{3}r^2 + \frac{2}{3}r^4\right)z^{-4} - \left(\frac{2}{5}r + \frac{4}{3}r^3 + \frac{4}{15}r^5\right)z^{-5} + \cdots \Bigg) \tag{10.16}$$
$$\approx \left(\frac{2}{T}\right)^r \frac{A_n(z^{-1},r)}{A_n(z^{-1},-r)}$$

and

$$s^{-r} \approx \left(\frac{2}{T}\right)^{-r} \frac{A_n(z^{-1},-r)}{A_n(z^{-1},r)} \tag{10.17}$$

where $A_n(z^{-1}, r)$, and likewise $A_n(z^{-1}, -r)$, can be found by the following iterations:

$$A_0(z^{-1}, r) \tag{10.18}$$
$$A_n(z^{-1}, r) = A_{n-1}(z^{-1}, r) - c_n z^{-n} A_{n-1}(z, r)$$

with the coefficient:

$$c_n = \begin{cases} r/n, & n \text{ is odd} \\ 0, & n \text{ is even} \end{cases}$$

Without doubt the accuracy of the approximation is determined by the selection of n.

10.2.4.2   IMPLEMENTATION OF FRACTIONAL-ORDER PID CONTROLLERS

By substituting the above discrete-time approximation of the fractional-order Laplace operator s^r and s^{-r} into Eq. (10.11), the fractional-order PID controller in Eq. (10.10) can be approximated in discrete-time domain as:

$$u(k) = K_P\, e(k) + K_{I,\lambda}\left(\frac{2}{T}\right)^{-\lambda} \frac{A_n(z^{-1},-\lambda)}{A_n(z^{-1},\lambda)} e(k) + K_{D,\mu}\left(\frac{2}{T}\right)^{\mu} \frac{A_n(z^{-1},\mu)}{A_n(z^{-1},-\mu)} e(k)$$

$$(10.19)$$

where the index k stands for the k-th sampling time, e.g., $e(k) = e(t = kT_s)$ with sampling time T_s.

10.3   AN UNDERACTUATED SYSTEM: A ROTARY PENDULUM SYSTEM

A rotary pendulum system (Figure 10.1) which uses a permanent magnet DC motor to drive a two-link rotary pendulum is the underactuated system employed to demonstrate the performance of the fractional-order PID controller in this research. The rotary pendulum system is an underactuated system because there is only one actuator to drive link 1 but no actuator to drive link 2. Both links have one rotational degree of freedom (d.o.f.) around different axes. The motion of link 2 is coupled with link 1 so it has to be considered when controlling link 1.

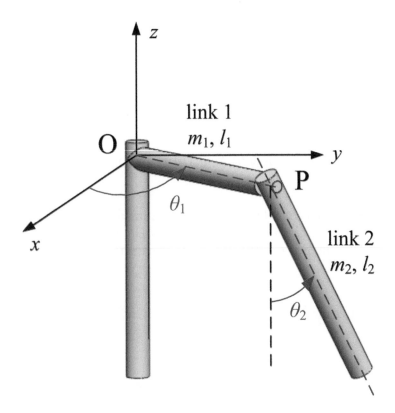

FIGURE 10.1   The two-link rotary pendulum.

10.3.1   DC MOTOR MODEL

A DC motor model is used to drive the two-link rotary pendulum. The armature voltage is e_a, and the current through the armature is i. The armature current generates a motor torque $\tau_M = K_T i$, where K_T is the torque constant of the DC motor. Assume that the angular displacement of the motor shaft is θ and the load is τ_L, then the mechanical dynamics of the DC motor is:

$$J_M \frac{d^2\theta}{dt^2} + B_M \frac{d\theta}{dt} + \tau_L = K_T i \qquad (10.20)$$

where J_M and B_M are the rotational inertia and the rotational damping coefficient of the motor, respectively.

As the motor shaft rotates at a speed $\omega = d\theta/dt$, the so-called back e.m.f. is induced by the relationship $e_{b.e.m.f.} = K_E \omega$ where K_E is the back e.m.f. constant of the DC motor and $K_E = K_T$ in the SI unit system. Then applying the Kirchhoff's voltage law to the armature gives the electrical dynamics of the DC motor below:

$$L_a \frac{di}{dt} + R_a i + K_E \omega = e_a \tag{10.21}$$

where L_a and R_a are the inductance and resistance of the armature, respectively.

10.3.2 TWO-LINK ROTARY PENDULUM MODEL

The mechanical part of the rotary pendulum, driven by the torque τ_L, consists of two rotational links of 1 d.o.f. each: link 1 and link 2 as shown in Figure 10.1. Link 1 is connected to the motor shaft and rotates around the center O, and link 2 is connected to the other end of link 1 and rotates around point P in a plane normal to link 1. The masses of link 1 and 2 are m_1 and m_2, respectively. The lengths of link 1 and 2 are l_1 and l_2, respectively. And the moment of inertia of link 1 around O is J_1. Assume that θ_1 and θ_2 are the angular displacements of link 1 and 2, respectively. Then the angular velocities of link 1 and 2 are $\dot{\theta}_1$ and $\dot{\theta}_2$ respectively. Since the center O is on the motor shaft, it is straightforward $\theta_1 = \theta$.

The dynamic model of the two-link rotary pendulum can be derived using the Euler-Lagrange equation [10]:

$$\frac{d}{dt}\left(\frac{\partial L}{\partial \dot{q}_i}\right) - \frac{\partial L}{\partial q_i} = Q_{nc,i} \tag{10.22}$$

where the Lagrangian L is the difference of the kinetic energy T and potential energy V of the system ($L = T - V$), q_i is the i-th generalized coordinate, and $Q_{nc,i}$ is the i-th non-conservative generalized force. The kinetic energy is:

$$T = \frac{1}{2}J_1\dot{\theta}_1^2 + \frac{1}{2}m_2l_1^2\dot{\theta}_1^2 + \frac{1}{6}m_2l_2^2\sin^2\theta_2\dot{\theta}_1^2 + \frac{1}{6}m_2l_2^2\dot{\theta}_2^2 + \frac{1}{2}m_2l_1l_2\cos\theta_2\dot{\theta}_1\dot{\theta}_2$$

$$(10.23)$$

where only the first term accounts for link 1, and the other 4 terms are from link 2. Meanwhile, the potential energy only comes from link 2, which is:

$$V = -m_2l_2\,g\,\cos\theta_2 \tag{10.24}$$

Therefore the Lagrangian is:

$$L = \frac{1}{2}J_1\dot{\theta}_1^2 + \frac{1}{2}m_2l_1^2\dot{\theta}_1^2 + \frac{1}{6}m_2l_2^2\sin^2\theta_2\dot{\theta}_1^2 + \frac{1}{6}m_2l_2^2\dot{\theta}_2^2 + \frac{1}{2}m_2l_1l_2\cos\theta_2\dot{\theta}_1\dot{\theta}_2 + m_2l_2g\cos\theta_2$$

$$(10.25)$$

where θ_1 and θ_2 are chosen as the 2 generalized coordinates, then the corresponding Euler-Lagrange equations are:

$$\frac{d}{dt}\left(\frac{\partial L}{\partial \dot{\theta}_1}\right) - \frac{\partial L}{\partial \theta_1} = \tau_L \tag{10.26}$$

and

$$\frac{d}{dt}\left(\frac{\partial L}{\partial \dot{\theta}_2}\right) - \frac{\partial L}{\partial \theta_2} = 0 \tag{10.27}$$

The dynamic equation of the mechanical subsystem then can be expressed as follows:

$$J_{11}\ddot{\theta}_1 + J_{12}\ddot{\theta}_2 + \frac{2}{3}m_2l_2^2\sin\theta_2\cos\theta_2\dot{\theta}_1\dot{\theta}_2 - \frac{1}{2}m_2l_1l_2\sin\theta_2\dot{\theta}_2^2 = \tau_L$$

$$(10.28)$$

$$J_{12}\ddot{\theta}_1 + J_{22}\ddot{\theta}_2 - \frac{1}{3}m_2l_2^2\sin\theta_2\cos\theta_2\dot{\theta}_1^2 + m_2l_2g\sin\theta_2 = 0 \tag{10.29}$$

where

$$J_{11} = J_1 + J_m + m_2 l_1^2 + \frac{1}{3} m_2 l_2^2 \sin^2 \theta_2, \quad J_{12} = \frac{1}{2} m_2 l_1 l_2 \cos \theta_2, \quad \text{and} \quad J_{22} = \frac{1}{3} m_2 l_2^2.$$

10.3.3 COMPLETE SYSTEM MODEL

The complete system consists of a DC motor represented by Eq. (10.20) and (10.21), and a two-link rotary pendulum represented by Eq. (10.28) and (10.29). Since Eq. (10.28) gives the term τ_L, it can be substituted into Eq. (10.20), and using the fact that $\theta_1 = \theta$, then the whole system therewith can be represented by the dynamic model below.

$$L_a \frac{di}{dt} + R_a i + K_E \frac{d\theta_1}{dt} = e_a$$

$$(J_M + J_{11}) \frac{d^2 \theta_1}{dt^2} + J_{12} \frac{d^2 \theta_2}{dt^2} \tag{10.30}$$

$$+ \frac{2}{3} m_2 l_2^2 \sin \theta_2 \cos \theta_2 \left(\frac{d\theta_1}{dt} \right) \cdot \left(\frac{d\theta_2}{dt} \right) - \frac{1}{2} m_2 l_1 l_2 \sin \theta_2 \left(\frac{d\theta_2}{dt} \right)^2 + B_M \left(\frac{d\theta_1}{dt} \right) = K_T i \tag{10.31}$$

$$J_{12} \frac{d^2 \theta_1}{dt^2} + J_{22} \frac{d^2 \theta_2}{dt^2} - \frac{1}{3} m_2 l_2^2 \sin \theta_2 \cos \theta_2 \left(\frac{d\theta_1}{dt} \right)^2 + m_2 l_2 g \sin \theta_2 = 0 \tag{10.32}$$

10.4 ESTIMATION OF CONTROLLER PARAMETERS

The aforementioned fractional-order Laplace operator approximation is used to design the fractional-order PID controller for the rotary pendulum system, and the proposed fractional-order Ziegler-Nichols method is employed to tune parameters. Then the results will be compared with the result of using the traditional Ziegler-Nichols method tuned integer-order PID controller. One important step of the Ziegler-Nichols method is to find the ultimate gain K_u and the corresponding period T_u for consistent oscillation. However, the trial-and-error approach would be time-consuming; it is desired to have the theoretical groundwork to make a reasonable estimation of K_u to save trouble.

The Ruth-Hurwitz stability criterion [11] can be used to find the ultimate gain K_u in a linear system. Since the rotary pendulum system is nonlinear, it helps if the local linearization method is used to linearize the system around an operating point before using the Ruth-Hurwitz stability criterion to find the ultimate gain K_u. The approach will be detailed in this section.

10.4.1   LOCAL LINEARIZATION OF THE TWO-LINK ROTARY PENDULUM SYSTEM

The equilibrium point of the rotary pendulum, i.e., $\theta_2 = 0$, is chosen to be the operating point for local linearization. First assume that $\theta_2 \approx 0$. Next linearize the dynamic Eqs. (10.31)–(10.32) by using the approximation $\sin \theta_2 \approx \theta_2$, $\cos \theta_2 \approx 1$, $\dot{\theta}_2 \approx 0$, and at equilibrium, then setting the higher order terms to zeros. Hence, the linearized model of the rotary pendulum can be expressed as:

$$L_a \frac{di}{dt} + R_a i + K_E \frac{d\theta_1}{dt} = e_a$$

$$(10.33)$$

$$\hat{J}_{11} \frac{d^2\theta_1}{dt^2} + \hat{J}_{12} \frac{d^2\theta_2}{dt^2} + B_M \frac{d\theta_1}{dt} = K_T i$$

$$(10.34)$$

$$\hat{J}_{12} \frac{d^2\theta_1}{dt^2} + \hat{J}_{22} \frac{d^2\theta_2}{dt^2} + m_2 l_2 g \theta_2 = 0$$

$$(10.35)$$

where $\hat{J}_{11} = J_M + J_1 + m_2 l_1^2$, $\hat{J}_{12} = \frac{1}{2} m_2 l_1 l_2$, and $\hat{J}_{22} = \frac{1}{3} m_2 l_2^2$.

Taking the Laplace transform of Eqs. (10.33)–(10.35) yields the following system of equations:

$$L_a s I + R_a I + K_E s \Theta_1 = E_a$$

$$(10.36)$$

$$\hat{J}_{11}s^2\Theta_1 + \hat{J}_{12}s^2\Theta_2 + B_M s\Theta_1 = K_T I \tag{10.37}$$

$$\hat{J}_{12}s^2\Theta_1 + \hat{J}_{22}s^2\Theta_2 + m_2 l_2 g\Theta_2 = 0 \tag{10.38}$$

Since Eq. (10.38) contains only two unknowns Θ_1 and Θ_2, Θ_2 can be expressed in terms of Θ_1 as:

$$\Theta_2 = -\frac{\hat{J}_{12}s^2}{\hat{J}_{22}s^2 + m_2 l_2 g}\Theta_1 \tag{10.39}$$

By substituting the result into Eq. (10.37), Θ_1 can be expressed in terms of I as:

$$\Theta_1 = \frac{K_T\left(\hat{J}_{22}s^2 + m_2 l_2 g\right)}{\left(\hat{J}_{11}s^2 + B_M s\right)\left(\hat{J}_{22}s^2 + m_2 l_2 g\right) - \hat{J}_{12}^2 s^4} I \tag{10.40}$$

and the transfer function of link 1's angular displacement to armature current is:

$$\frac{\Theta_1}{I} = \frac{K_T\left(\hat{J}_{22}s^2 + m_2 l_2 g\right)}{\left(\hat{J}_{11}s^2 + B_M s\right)\left(\hat{J}_{22}s^2 + m_2 l_2 g\right) - \hat{J}_{12}^2 s^4} \tag{10.41}$$

Alternatively, Eq. (10.40) can be rewritten in a different way as:

$$I = \frac{\left(\hat{J}_{11}s^2 + B_M s\right)\left(\hat{J}_{22}s^2 + m_2 l_2 g\right) - \hat{J}_{12}^2 s^4}{K_T\left(\hat{J}_{22}s^2 + m_2 l_2 g\right)}\Theta_1 \tag{10.42}$$

Then by substituting Eq. (10.42) into Eq. (10.36), Θ_1 can be expressed in terms of E_a as:

$$\Theta_1 = \frac{K_T\left(\hat{J}_{22}s^2 + m_2 l_2 g\right)}{\left(L_a s + R_a\right)\left(\left(\hat{J}_{11}s^2 + B_M s\right)\left(\hat{J}_{22}s^2 + m_2 l_2 g\right) - \hat{J}_{12}^2 s^4\right) + K_T K_E s\left(\hat{J}_{22}s^2 + m_2 l_2 g\right)} E_a \tag{10.43}$$

And the transfer function of link 1's angular displacement to armature voltage can be expressed as:

$$\frac{\Theta_1}{E_a} = \frac{K_T(\hat{J}_{22}s^2 + m_2l_2g)}{a_0s^5 + a_1s^4 + a_2s^3 + a_3s^2 + a_4s} \tag{10.44}$$

where

$$a_0 = L_a(\hat{J}_{11}\hat{J}_{22} - \hat{J}_{12}{}^2)$$

$$a_1 = R_a(\hat{J}_{11}\hat{J}_{22} - \hat{J}_{12}{}^2) + L_a B_M \hat{J}_{22}$$

$$a_2 = L_a \hat{J}_{11}m_2l_2g + R_a B_M \hat{J}_{22} + K_T K_E \hat{J}_{22}$$

$$a_3 = L_a B_M m_2l_2g + R_a \hat{J}_{11}m_2l_2g$$

$$a_4 = R_a B_M m_2l_2g + K_T K_E m_2l_2g \qquad .$$

10.4.2 TRANSFER FUNCTION OF THE CLOSED-LOOP CONTROLLED ROTARY PENDULUM SYSTEM

Figure 10.2 shows the block diagram of a typical closed-loop control system. The plant is the DC motor-driven rotary pendulum system, the reference input r is the desired link 1's angular displacement input θ_{1d}, the system output y is the actual link 1's angular displacement output θ_{1a}, and the control effort u is the armature voltage e_a in this research. When using a P controller with gain K_p to control the rotary pendulum system, the closed-loop transfer function of the actual angular displacement output θ_{1a} to the desired angular displacement input θ_{1d} becomes:

$$\frac{\Theta_{1a}}{\Theta_{1d}} = \frac{\dfrac{K_p\Theta_1}{E}}{1 + \dfrac{K_p\Theta_1}{E}} = \frac{K_p(k_1s^2 + k_2)}{a_0s^5 + a_1s^4 + a_2s^3 + (a_3 + k_1K_p)s^2 + a_4s + k_2K_p} \tag{10.45}$$

where $k_1 = K_T\hat{J}_{22}$ and $k_2 = K_T m_2l_2g$. And its characteristic equation is:

$$a_0s^5 + a_1s^4 + a_2s^3 + (a_2s^3 + k_1K_p)s^2 + a_4s + k_2K_p = 0 \tag{10.46}$$

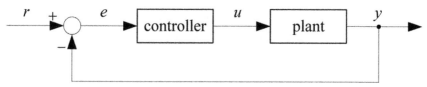

FIGURE 10.2 Block diagram of a closed-loop control system.

10.4.3 ESTIMATION OF THE ULTIMATE GAIN USING THE RUTH TABLE

The following DC motor parameters and two-link rotary pendulum parameters are used: $L_a = 4.6909 \times 10^{-3}$, $R_a = 2.5604\ \Omega$, $K_T = 1.8259 \times 10^{-2}\ \text{N·m/A}$, $K_E = 1.8259 \times 10^{-2}\ \text{V/(rad/s)}$, $J_M = 7.2019 \times 10^{-5}\ \text{kg·m}^2$, $B_M = 9.1358 \times 10^{-5}\ \text{N·m(rad/s)}$, m_1 0.056 kg, $l_1 = 0.16$ m, J_1 0.001569 kg·m², $m_2 = 0.022$kg, and $l_2 = 0.16$m. The coefficients of the characteristic equation calculated are $a_0 \approx 1.5691 \times 10^{-9}$, $a_1 \approx 8.5655 \times 10^{-7}$, $a_2 \approx 4.6355 \times 10^{-7}$, $a_3 \approx 1.949 \times 10^{-4}$, $a_4 \approx 1.959 \times 10^{-5}$, $k_1 \approx 3.4278 \times 10^{-6}$, and $k_2 \approx 6.3051 \times 10^{-4}$. And the Ruth table [11] is shown in Table 10.1.

TABLE 10.1   Ruth Table with K_P as a Variable

s^5	1.5691×10^{-9}	4.6355×10^{-7}	1.959×10^{-5}
s^4	8.5655×10^{-7}	$(1.949 \times 10^{-4} + 3.4278 \times 10^{-6} K_P)$	$6.3051 \times 10^{-4} K_P$
s^3	A_1	A_2	–
s^2	B_1	$6.3051 \times 10^{-4} K_P$	–
s^1	C_1	–	–
s^0	$6.3051 \times 10^{-4} K_P$	–	–

where , $A_1 = \dfrac{-5.3787 \times 10^{-15} K_P + 9.123 \times 10^{-14}}{8.5655 \times 10^{-7}}$

$$A_2 = \frac{-2.2512 \times 10^{-10} K_P + 3.818 \times 10^{-9}}{3.4278 \times 10^{-6} K_P + 1.949 \times 10^{-4}}$$

,

$$B_1 = \frac{-6.32 \times 10^{-26} K_P^3 - 6.1148 \times 10^{-24} K_P^2 + 8.275 \times 10^{-23} K_P + 6.642 \times 10^{-22}}{-1.8437 \times 10^{-20} K_P^2 - 7.3558 \times 10^{-19} K_P + 1.7781 \times 10^{-17}}$$

$$C_1 = \frac{C_n}{C_d}$$

$$C_n = (-2.1433 \times 10^{-43} K_P^5 - 4.9155 \times 10^{-42} K_P^4 + 1.0447 \times 10^{-39} K_P^3$$

$$- 1.9461 \times 10^{-38} K_P^2 - 5.67889 \times 10^{-38} K_P + 2.1721 \times 10^{-36})$$

$$C_d = (-1.8556 \times 10^{-37} K_P^4 - 2.8504 \times 10^{-35} K_P^3 - 7.7785 \times 10^{-34} K_P^2$$

$$+ 1.5764 \times 10^{-32} K_P + 1.1088 \times 10^{-31})$$

　　There are three cases in determining the ultimate gain K_u: (1) $A_1 = 0$, (2) B_1 = 0, and (3) $C_1 = 0$. However, only the third case gives a reasonable solution. For the third case, i.e., $C_1 = 0$ or the first term in the s^1 row is zero, $K_P \approx 16.9592$ is found which gives $A_1 > 0$, $B_1 > 0$, and is chosen as the ultimate gain K_u. (The corresponding Ruth table is shown in Table 10.2.) Though the ultimate gain $K_u \approx 16.9592$ is calculated using the linearized model, not the actual nonlinear model, it's good to choose it as an initial guess to find the closest K_u and T_u, and therewith determine the corresponding controller parameters (which will be given in the next section) to complete the controller design.

TABLE 10.2　　Ruth Table with $K_P \approx 16.9592$

s^5	1.5691×10^{-9}	4.6355×10^{-7}	1.959×10^{-5}
s^4	8.5655×10^{-7}	2.5303×10^{-4}	0.0107
s^3	1.2805×10^{-11}	6.3217×10^{-10}	–
s^2	2.1075×10^{-4}	0.0107	–
s^1	0	–	–
s^0	0.0107	–	–

10.5　SIMULATION RESULTS

This research will evaluate the fractional-order PID controller's performance by simulation in this section. The point-to-point maneuver is the task. To maneuver a system to a target position, step input is often used in textbooks to evaluate the controller's performance by the response. It's also used traditionally in the Ziegler-Nichols method in determining the PID parameters. Yet trajectory planning makes the transition smoother and is practically used in motion control. Trapezoidal curve (T curve) and sigmoid curve (S curve) are two popularly used speed profiles in trajectory planning. For simplicity and demonstration purpose, a symmetric T-curve speed profile is adopted as $\dot{\theta}_{1d}$ to generate the desired link 1's angular displacement trajectory θ_{1d} (as shown in Figure 10.2). And the planned trajectory serves for finding the ultimate gain K_u and the ultimate period T_u using Ziegler-Nichols method, and evaluating the performance of the tuned PID controller and fractional-order PID controllers of 3 different fractions in this section.

10.5.1 TRAJECTORY PLANNING

A symmetric T-curve speed profile is adopted as the rotational speed and used for generating the desired angular displacement trajectory. Basically, it consists of one constant acceleration phase, one constant speed phase, and one constant deceleration phase. Assume the angular acceleration is α_m at the first phase ($0 \le t < t_1$), the angular speed $\omega_m = \alpha_m t_1$ at the second phase ($t_1 \le t < t_2$), and the angular deceleration $-\alpha_m$ at the third phase ($t_2 \le t < t_3$). After the third phase ($t \ge t_3$), the desired position stays at the target position (total angular displacement). The desired angular acceleration $\alpha_d(t) = \ddot{\theta}_d(t)$, angular speed $\omega_d(t) = \dot{\theta}_d(t)$, and angular displacement $\theta_d(t)$ of a general symmetric T-curve angular speed (Figure 10.3) can be described as follows:

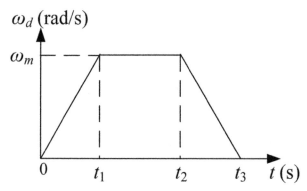

FIGURE 10.3 Trapezoidal curve angular speed profile.

1. $0 \le t < t_1$

$$\begin{cases} \alpha_d = \alpha_m = \dfrac{\omega_m}{t_1} \\[2mm] \omega_d = \alpha_m t \\[2mm] \theta_d = \dfrac{1}{2}\alpha_m t^2 \end{cases}$$

(10.47)

2. $t_1 \le t < t_2$

$$\begin{cases} \alpha_d = 0 \\[2mm] \omega_d = \omega_m \\[2mm] \theta_d = \dfrac{1}{2}\omega_m t_1 + \omega_m(t - t_1) \end{cases}$$

(10.48)

3. $t_2 \le t < t_1$

$$\begin{cases} \alpha_d = -\alpha_m \\[2mm] \omega_d = \omega_m - \alpha_m(t - t_2) \\[2mm] \theta_d = \dfrac{1}{2}\omega_m t_1 + \omega_m(t_2 - t_1) + \omega_m(t - t_2) - \dfrac{1}{2}\alpha_m(t - t_2)^2 \end{cases}$$

(10.49)

4. $t \ge t_3$

$$\begin{cases} \alpha_d = 0 \\[2mm] \omega_d = 0 \\[2mm] \theta_d = \omega_m t_2 \end{cases}$$

(10.50)

10.5.2   RESULTS AND DISCUSSIONS

This section presents simulation results of the DC motor-driven two-link rotary pendulum system. The parameters in Section 10.4.3 are used in this section. Initial values of $\theta_1 = 0$, $\theta_2 = 0$, $\dot\theta_1 = 0$, and $\dot\theta_2 = 0$ are chosen. The target link 1's angular displacement is set to be $\theta_{1d} = \pi/2 = 90°$. Hence the T-curve angular speed profile mentioned in the previous subsection is employed to generate link 1's trajectory with parameters $\dot\theta_{1,\max} = \pi/3$ rad/s, $t_1 = 0.5$ s, and $t_3 = 2$ s.

The first step of the Ziegler-Nichols method is to set $K_I = 0$, and $K_D = 0$, then find the ultimate gain K_u which makes the closed-loop system have consistent oscillation in the output. As suggested in Section 10.4.3, the theoretical value $K_u \approx 16.9592$ calculated from the Ruth table built for the linearized model is used as an initial guess to find the ultimate gain of the actual nonlinear model. Within a few trials, it is found that the closed-loop system have consistent oscillation in the output when $K_p = 16.96$ as shown in Figure 10.4, hence the ultimate gain K_u is chosen to be 16.96. Also, the ultimate period T_u is about 0.80736 s as seen in Figure 10.4.

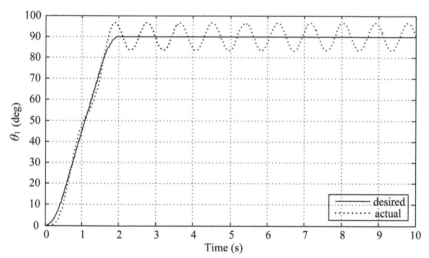

FIGURE 10.4   Link 1's response under P control ($K_p = 16.96$, $K_I = 0$, $K_D = 0$).

According to the Ziegler-Nichols method, choose the PID parameters to be $K_p = 10.176$, $K_I = 25.2081$, and $K_D = 1.02696$. The simulation results using the tuned PID parameters show that the maximum overshoot of link 1 is $5.135°$, and the settling time for the settling window of $\pm5\%$ ($\pm 4.5°$ for $90°$ motion) occurs at $t_s = 2.163$s (0.163 s after the 2-second T-curve motion) as shown in Figure 10.5. There is a trend of 1/4 decay oscillation after 2 s. The maximum swing angle of link 2 is $\theta_2 \approx -1.7°$ (Figure 10.6). And the time history of the control armature voltage is shown in Figure 10.7. The armature voltage also shows 1/4 decay oscillation after 2 s just like Figure 10.5.

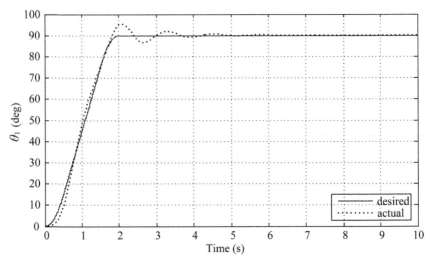

FIGURE 10.5   Link 1's response under PID control ($K_P = 10.176$, $K_I = 25.2081$, $K_D = 1.02696$).

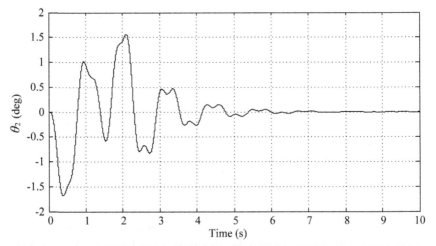

FIGURE 10.6   Link 2's response under PID control ($K_P = 10.176$, $K_I = 25.2081$, $K_D = 1.02696$).

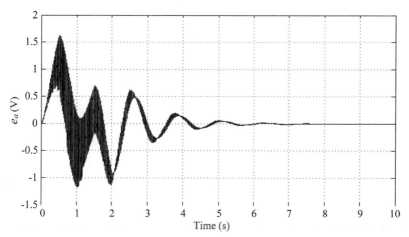

FIGURE 10.7   Armature voltage under PID control (K_P = 10.176, K_I = 25.2081, K_D = 1.02696).

Next the fractional-order PID control will be employed so as to make comparison with the tuned (intger order) PID control. As proposed in Subsection 10.2.2.2, $K_P = 0.6K_u$, $K_{I\lambda} = K_P / T_I^\lambda = K_P(2/T_u)^\lambda$, and $K_{D,\mu} = K_P T_D^\mu = K_P(T_u/8)^\mu$ are used for the fractional integral gain and the fractional derivative gain, respectively. Though the output θ_1 and the internal state θ_2 are coupled, there is a certain relationship between them. It would be straightforward to choose $\lambda = \mu$ then. Three sets of fractional-order PID parameters are used for comparison: (1) $\lambda = \mu = 0.4$, (2) $\lambda = \mu = 0.5$, and (3) $\lambda = \mu = 0.6$. Here are the results:

1. $\lambda = \mu = 0.4$: The controller parameters are chosen to be $K_P = 10.176$, $K_{I,\lambda} \approx 14.62723$, and $K_{D,\mu} \approx 4.0660$. As shown in Figure 10.8, the maximum overshoot of link 1 is about 0.71° and the actual angular displacement of link 1 stays within the settling window after the planned 2-sec. moving time.

    The maximum swing angle of link 2 is around –2.9° (Figure 10.9). And the time history of the control armature voltage is shown in Figure 10.10. However, there is no sign of decay after 2 s. After 2 s, Link 1 and link 2 are oscillating between ±1° and ±1.3°, respectively, and the armature voltage is oscillating between ±0.4° (V).

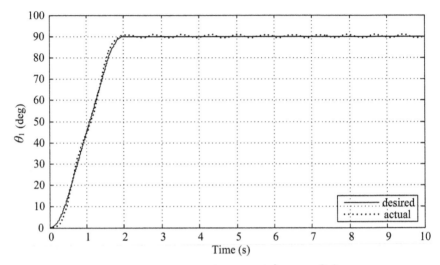

FIGURE 10.8   Link 1's response under PI$^\lambda$D$^\mu$ control ($\lambda = \mu = 0.4$).

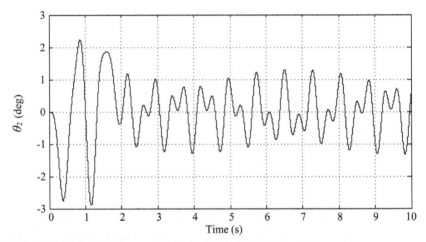

FIGURE 10.9   Link 2's response under PI$^\lambda$D$^\mu$ control ($\lambda = \mu = 0.4$).

2. $\lambda = \mu = 0.5$: The controller parameters are chosen to be $K_p = 10.176$, $K_{I,\lambda} \approx 16.01616$, and $K_{D,\mu} \approx 3.2327$. As shown in Figure 10.11, the maximum overshoot of link 1 is about 0.15° and the actual angular displacement of link 1 stays within the settling window after the planned 2-sec. moving time.

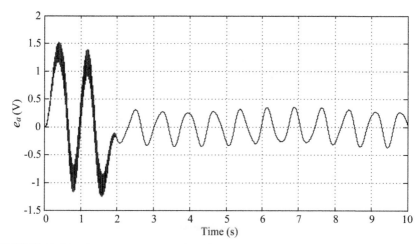

FIGURE 10.10 Armature voltage under PI$^\lambda$D$^\mu$ control ($\lambda = \mu = 0.4$).

The maximum swing angle of link 2 is around $-2.8°$ (Figure 10.12). And the time history of the control armature voltage is shown in Figure 10.13. Again, there is no sign of decay after 2 s. After 2 s, Link 1 and link 2 are oscillating between $\pm 0.5°$ and $\pm 0.8°$, respectively, and the armature voltage is oscillating between ± 0.15(V).

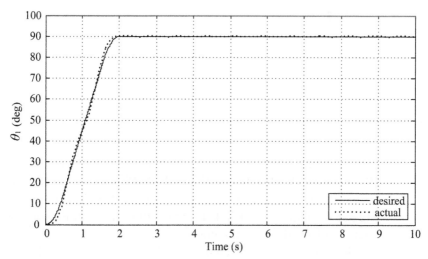

FIGURE 10.11 Link 1's response under PI$^\lambda$D$^\mu$ control ($\lambda = \mu = 0.5$).

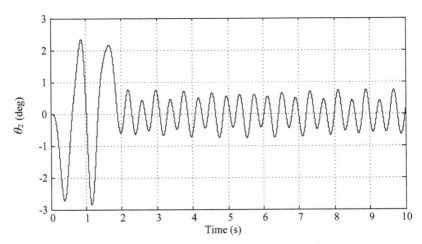

FIGURE 10.12   Link 2's response under PI$^\lambda$D$^\mu$ control ($\lambda = \mu = 0.5$).

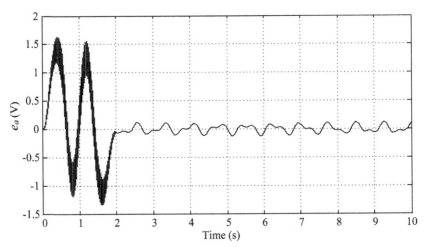

FIGURE 10.13   Armature voltage under PI$^\lambda$D$^\mu$ control ($\lambda = \mu = 0.5$).

(3) $\lambda = \mu = 0.6$: The controller parameters are chosen to be $K_P = 10.176$, $K_{I,\lambda} \approx 17.53698$, and $K_{D,\mu} \approx 2.57018$. As shown in Figure 10.14, the maximum overshoot of link 1 is about 0.4° and the actual angular displacement of link 1 stays within the settling window after the planned 2-sec. moving time.

The maximum swing angle of link 2 is around –2.8°(Figure 10.15). And the time history of the control armature voltage is shown in Figure 10.16. However, there is no sign of decay after 2 s. After 2 s, Link 1 and link 2 are oscillating between ± 1° and ± 0.9°, respectively, and the armature voltage is oscillating between ± 0.2 (V).

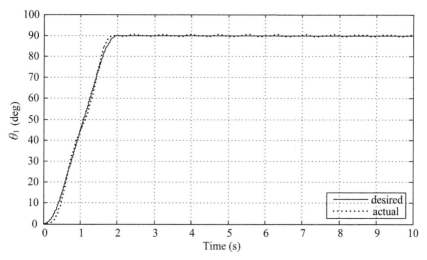

FIGURE 10.14   Link 1's response under PI$^\lambda$D$^\mu$ control ($\lambda = \mu = 0.6$).

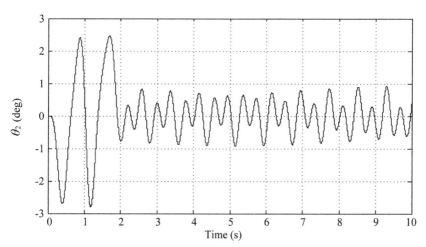

FIGURE 10.15   Link 2's response under PI$^\lambda$D$^\mu$ control ($\lambda = \mu = 0.6$).

Generally speaking, the performance of all three fractional-order PID controllers, designed using the proposed generalized Ziegler-Nichols method, is better than that of the integer-order PID controller, designed using the Ziegler-Nichols method in terms of tracking during 2-sec. motion, overshoot, and settling time. However, oscillation continues even after 10 sec. if any of these three fractional-order PID controllers is used, while oscillation decays to zero if the integer-order PID controller is used. This might suggest that the proposed generalized Ziegler-Nichols method should be further improved in order to have the system converges to zero.

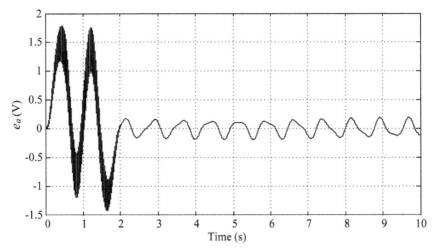

FIGURE 10.16   Armature voltage under $PI^\lambda D^\mu$ control ($\lambda = \mu = 0.6$).

Now among these three fractional-order PID controllers, the one comes with $\lambda = \mu = 0.5$ shows the best performance in every way which implies the system is more likely to have fractional order 0.5 than 0.4 or 0.6.

10.6   CONCLUSIONS

The performance of the tuned PID controller and three tuned $PI^\lambda D^\mu$ controllers are compared when applied to control an underactuated system. The use of local linearization and Ruth-Hurwitz stability criterion gives a very close approximation of the ultimate gain for the actual nonlinear system. It seems that λ and μ should be equal based on the order of the coupled dynamics of link 1 and link 2. A generalized Ziegler-Nichols method for the fractional-order PID controller is proposed to include the integer-order PID controller as a special case $\lambda = \mu = 1$. It keeps the simplicity of the Ziegler-Nichols method. The simulation results show that the system is more likely to have fractional order 0.5 than 0.4 or 0.6 based on their performance. Meanwhile, these three tuned $PI^{\lambda}D^{\mu}$ controllers work better in terms of tracking during 2-sec. motion, overshoot, and settling time, especially settling time, but not oscillation decay rate. Yet these results show the fractional-order controller is promising if the oscillation problem can be dissolved. Hence more vigorous researches should be conducted to improve the proposed generalized Ziegler-Nichols method so it can be used in versatile control applications.

KEYWORDS

- **Commande Robuste d'Ordre Non-Entier**
- **fractional-order**
- **parameter tuning**
- **trajectory planning**
- **underactuated system**
- **Ziegler-Nichols method**

REFERENCES

1. Miller, K. S., & Ross, B., (1993). *An Introduction to the Fractional Calculus and Fractional Differential Equations*. Wiley: New York.
2. Ziegler, J. G., & Nichols, N., (1942). Optimum settings for automatic controllers. *Transactions of the ASME, 64*, 759–768.
3. Monje, C. A., Vinagre, B. M., Feliu, V., & Chen, Y., (2008). Tuning and auto-tuning of fractional-order controllers for industry applications. *Control Engineering Practice, 16*(7), 798–812.
4. Valério, D., & Sá Da Costa, J., (2006). Tuning of fractional PID controllers with Ziegler-Nichols-type rules. *Science Signal Processing, 86*(10), 2771–2784.
5. Singh, A. P., Srivastava, T., Agrawal, H., & Srivastava, P., (2017). Fractional-order controller design and analysis for crane system. *Progress in Fractional Differentiation and Applications, 3*(2), 155–162. doi: 10.18576/pfda/030206.

6. Singh, A. P., Agarwal, H., & Srivastava, P., (2015). Fractional-order controller design for inverted pendulum on a cart system (POAC). *WSEAS Transactions on Systems and Control, 10,* 172–178.

7. Fan, H., Sun, Y., & Zhang, X., (2007). Research on fractional-order controller in servo press control system. *2007 IEEE International Conference on Mechatronics and Automation.* Harbin, China. [Online] doi: 10.1109/ICMA.2007.4304026. https://ieeexplore.ieee.org/document/4304026 (accessed on 13 May 2020).

8. Oustaloup, (1995). *La Derivation non-Entiere.* Hermes: Paris, France.

9. Chen, Y. Q., & Moore, K. L., (2002). Discretization schemes for fractional-order differentiators and integrators. *IEEE Transactions on Circuits and Systems I: Fundamental Theory and Applications, 49*(3), 363–367.

10. Meirovitch, L., (1970). *Methods of Analytical Dynamics.* McGraw-Hill: New York.

11. Nise, N. S., (2015). *Control Systems Engineering* (7th edn.). Wiley: New York.

CHAPTER 11

Evaluating the Effect of an Amputee's Physical Parameters of Pressure on a Lower-Limb Prosthetic Socket Using a Fuzzy-Logic-Based Model

VIMAL KUMAR PATHAK,[1] CHITRESH NAYAK,[2] and DEEPAK RAJENDRA UNUNE[3]

[1]Assistant Professor, Department of Mechanical Engineering, Manipal University Jaipur, Dehmi Kalan, Jaipur – 303007, Rajasthan, India, E-mail: vimalpthk@gmail.com

[2]Professor, Oriental Institute of Science and Technology, Bhopal, Madhya Pradesh, India

[3]Assistant Professor, Department of Mechanical Engineering, The LNM Institute of Technology, Jaipur, Rajasthan, India

11.1 INTRODUCTION

Over the past few years, the requirement for implants and medical devices in orthopedics has undertaken rapid growth due to some important factors including growing elderly population, technology developments, the rise in chronic diseases, and improved healthcare facilities in developing countries. Currently, there are more than 30 million people worldwide having amputations [1] and most of them involve the lower limb at the transtibial level [2]. With the help of prosthesis, amputees can improve the quality of life. The amputation rate in developing countries, including India, is about 45% of diabetic foot problems, with an estimated 50,000 amputations occurring per year [3]. The prosthetic socket act as a critical interface between amputation and residual limb, which is designed and developed in an iterative process by the prosthetist.

Regardless of the use of advanced technologies in socket manufacturing, definite stump-socket interaction takes place. This interaction results in excessive stresses, patient discomfort in wearing the prosthesis, pain, skin irritation, pistoning, and stump soft tissue damage [4]. It is accepted that the superiority of the socket fit is directly related to the pressure distribution produced at the residual limb-socket interface. The quality of the socket fit is considered to be good if the total load is supported by the pressure tolerant areas (e.g., areas of thick tissue) of the limb. The pressure-sensitive areas (e.g., areas where the bone is close to the surface) should be non-load bearing for a better comfort. Currently, it is forecasted that 20 present of prosthetic sockets fabricated are either discarded due to improper fitting or significantly require modification to ensure satisfactory 'fit,' signifying knowledge gap in procedures. The amputee discomfort and pain occurs due to high-pressure interaction at the stump-socket interface, which is one of the important factors to be considered in the area of prosthetics and orthotics. For the reason above, the determination of pressure distribution is useful in effective socket design and development. Additionally, the determination of pressure distribution at the stump-socket interface proves to be an effective parameter to enhance the quality of the socket fit [5].

The transducers have been used for pressure assessment purposes since late 1970 [6]. Several researchers have studied and investigated about force, pressure, displacement, strain, normal, and shear stress at the stump-socket interface [4, 7] using a diaphragm strain gauge, piston-type strain gauge, capacitive, piezoresistive based sensors [8, 9]. Most of the investigators prefer to use piezoresistive sensors such as force sensing resistors (FSR) due to their eminent features including the small size with a simple structure, thin construction, adequate flexibility, good sensitivity, and ease of use [10, 11]. In comparison to other sensors that could be either positioned within prosthetic sockets or mounted on socket wall, all piezoresistive sensors are very thin sheets; ideal to be positioned in-situ inside the prosthetic socket. The accurate pressure measurement required a suitable measurement technique, appropriate sensor; correct positioning of the sensor at the stump-socket interface. A suitable pressure measurement system should be able to produce actual results without changing the initial stump-socket interface condition. Pressure measurement help in the realization of the intricate problems confronted during a socket fitting.

Presently researchers are more focused on determining novel bio-materials for socket manufacturing and further integration of advanced tools like reverse engineering, CAD, and FEM to produce accurate virtual model development and pressure measurement [12–14]. The past research concluded that transducers are applicable for gathering pressure data in limited regions within the prosthetic socket. FEM requires detailed information of the patient's limb and socket geometry and material properties that is not available for each patient. The aforementioned studies have provided results that deal with the aid in understanding the critical issues faced in socket fitting. It is also found that advanced tools are needed to overcome the limitations of the traditional socket fitting process, providing practical results helpful for clinical significance [15, 16].

In presently available methods of pressure measurement, the pressure can be measured either by sensor introduction into the socket/limb interface, which affects the results collected or by altering the socket to insert the pressure transducer, leading to difficulties in daily use. For overcoming these drawbacks, a computational approach will be beneficial in predicting the pressure at the socket interface. For this reason, a fuzzy logic-based artificial intelligence model have been proposed for the determination of pressure under different conditions, i.e., static, and dynamic. In the field of prosthetics and orthotics, no work is available on applying the fuzzy logic model in pressure measurement and evaluation at the socket interface.

Soft computing techniques have the ability to describe non-trivial complex problems where input and output relations are non-linear. Soft computing techniques such as Fuzzy logic, ANN, evolutionary, and nature-inspired algorithms provide an adequate solution to the variety of complex problems, while acknowledging the uncertainty involved in the problem [17]. Fuzzy logic is effective when an accurate mathematical model is not available, can work with imprecise inputs, and at last can handle the non-linearity with ease. Therefore, in the present study, the authors have selected fuzzy logic technique to model pressure measured at different specific regions in terms of amputee physical parameters. The capability to model the relations amid the different loading conditions and pressure effects on the residual limb would provide an enhanced tool for examination and aid clinicians to analyze socket discomfort issues. It would also enable research on socket interface design and material and will help clinicians in advising involvements for amputees with complex residual limbs. Determining the influence of amputees' physical parameters on real-life pressure values at the socket will aid the prosthesis to better design ensuring the comfort of the amputees.

11.2   MATERIALS AND METHODS

Ten unilateral below-knee male amputees have been selected for the acquisition of pressure data for this investigation. The details of the amputees are given in Table 11.1. This study considered clinically in different cases. The selected amputees regularly used a prosthesis patella tendon bearing (PTB) socket with a uniform thickness of 5 mm with cotton liner. They had been using exo-skeletal transtibial prosthesis from the last 3 to 21 years with PTB socket manufacturing in Bhagwan Mahaveer Viklang Sahayata Samiti (BMVSS), Jaipur, India. The measurements have been carried out using the Flexi-Force sensor as shown in Figure 11.1.

TABLE 11.1   General Information about Patients

Patient	Age	Height (cm)	Weight (kg)	Using Prosthesis (Year)	Type of Amputee	Clinically Significant Cases	Stump Length (cm)
P1	30	172	66	4	Unilateral	Accident	Right 18.1 (±2.0)
P2	42	173	70	3	Unilateral	Trauma-vascular disease Short)	Left 15 (±2.0)
P3	36	174	72	3	Unilateral	Accident (Long)	Left 28 (±2.0)
P4	56	171	68	10	Unilateral	Infection	Left 16 (±2.0)
P5	70	168	63	9	Unilateral	Accident	Right 23 (±2.0)
P6	40	166	70	21	Unilateral	Conjugation	Right 22.8 (±2.0)
P7	46	162	63	8	Unilateral	Diabetic	Right 20.3 (± 2.0)
P8	26	168	52	12	Unilateral	Accident	Left 25 (± 2.0)
P9	50	167	64	15	Unilateral	Accident	Left 19.1 (±2.0)
P10	38	170	58	7	Unilateral	Diabetic	Right 19.3 (±2.0)

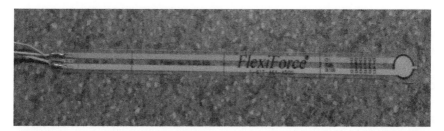

FIGURE 11.1   Flexi-force pressure sensor.

The characteristics of the flexi-force sensor are listed in Table 11.2. The sensor is of small thickness, flexible printed circuit, lightweight custom shape, and size. It can measure the force between any two contacting surfaces and is durable enough to stand up in most environments with a force range (0 to 445 N). These sensors can be easily integrated into the stump-socket interface. The sensor measure both static and dynamic forces between stump and socket. It is constructed from two layers of the substrate; this substrate is composed of polyester film (or polyimide in the case of the high-temperature sensors). On each layer, a conductive material (silver) is applied, followed by a layer of pressure-sensitive ink sensor. The adhesive is then used to laminate the two layers of substrate together to form the sensor. The sensor acts as a variable resistor in an electrical circuit. When the sensor is unloaded, its resistance is very high (greater than 5 MΩ); when a force is applied to the sensor, the resistance decreases.

TABLE 11.2   Characteristics of Flexi-Force Standard Model A201

Sl. No.	Parameters	Value	Unit
1.	Thickness	0.208	mm
2.	Length	197	mm
3.	Width	14	mm
4.	Sensing Area	9.53	mm^2
5.	Force Range	0–445	N

11.3   EXPERIMENTATION AND DATA ACQUISITION (DAQ)

The experiments have been carried out on the amputees using an experimental setup as shown in Figure 11.2. One step-down transformer is used to convert an AC voltage of 220 V to 9–0–9 V. Analogue to digital converter (ADC) is used to converts an analog signal to digital. The result is a sequence of digital values that have been converted from a continuous-time and continuous-amplitude analog signal to a discrete-time and discrete-amplitude digital signal.

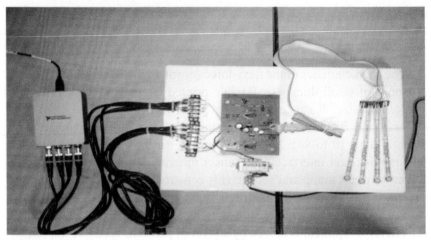

FIGURE 11.2   Flexi-force set up with a national instrument system.

One capacitor is used to stabilize the DC voltage. Resistance is used to drop the voltage. Then the experimental setup is connected to data acquisition (DAQ) block/card (DAQ-9171), four channels NI 9234 chassis (national instruments) by BNC probe cable to the virtual instrumentation software (Lab-view) for DAQ. DAQ is the process of measuring an electrical or physical phenomenon such as voltage, current, temperature, pressure, or sound with a computer. A DAQ system consists of sensors, DAQ measurement hardware, and a computer with programmable software. Compared to traditional measurement systems, PC-based DAQ systems exploit the processing power, productivity, display, and connectivity capabilities of industry-standard computers providing a more powerful, flexible, and cost-effective measurement solution. A Flexi-Force sensor is attached to the curved portion of the plastic frame, and the accelerometer board is mounted on its base of the oscillating mass. Flexible force sensor (FFS) system is ready to plug-n-play. The FFS works like any other bridge transducer, by converting non-linear resistance changes to a linear output voltage proportional to force. The calibration of FFS is performed by using the dead weights following standard calibration procedure. The accuracy and repeatability of measurements of FFS were analyzed. The static and dynamic loading tests were performed to obtain pressure sensing parameters. The linearity, repeatability, and hysteresis were found to be +/–4.5%, +/–4.1%, +/–6.1%. The drift was found to less than 5.6%.

In the current study, eight specific regions, as shown in Figure 11.3, have been identified to measure pressure at different loading conditions (half, full, and walking). The regions are Lateral Tibia (P1), Gastrocnemius (P2), PTB (P3), Kick Point (P4), Medial Tibia (P5), Medial Gastrocnemius (P6), Popliteal Depression (P7), and Lateral Gastrocnemius (P8). The interface pressure values have been recorded for two standing, viz. half body weight, and full body weight, for 50 seconds. For walking conditions, the amputee is asked to walk for 12 meters distance. Before initiating a measurement, all hardware components of the FFS system (socket, connecting the cable, converter) must be appropriately connected. The FFS is placed between the liner and socket at eight specific regions. Further, a pressure measurement at all eight regions can be viewed simultaneously in real-time using the software on a laptop/PC screen, and the measurement can be repeated if required.

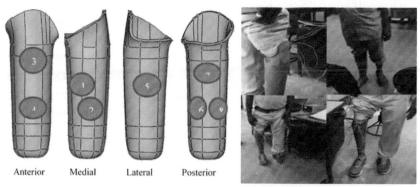

Anterior   Medial   Lateral   Posterior

FIGURE 11.3   The pressure points (Left) and fitting of the sensor (Right) on the limb.

The measured values of pressure at the eight specific regions are presented in Table 11.3. For this study, three trials were performed, and an average of the pressure data is selected and reported. The maximum pressure at all the three (walking, full, and half) conditions are shown in bold. From Table 11.3, it has observed that the strongest impact of maximum pressure between stump-socket interfaces is on the PTB.

TABLE 11.3　Static and Dynamic Pressure (KPa) Data Using Flexi-Force Sensor

Patient	Loading Condition	Pressure Points							
		P1	P2	P3	P4	P5	P6	P7	P8
Patient 1	(W)	161.4	206.5	278.5	242	193.4	165.12	179.6	179.4
	(F)	110.2	184.81	223.1	202.8	152.23	151.12	196	148.6
	(H)	24	39	55.2	46	36	33.2	49	29
Patient 2	(W)	79.4	198.4	315.2	222.5	146.4	210	204.4	210.6
	(F)	56	98.8	267	196	121.6	200.2	183	164.6
	(H)	24	42.4	65	48	41	35	42.8	35
Patient 3	(W)	125.6	111.4	239.5	148.8	166.4	79.6	153.6	86
	(F)	108.2	65.8	185	120	63	72	135.6	75.4
	(H)	29	24.2	49	34.5	24	25	45	20
Patient 4	(W)	108	167.4	271.6	229.6	136	204	220.6	224.4
	(F)	135.2	171.4	259	156	108	116	176	210.8
	(H)	27.2	40.4	36.3	66.3	41	36	27	28
Patient 5	(W)	102	127	246	176	243	100	184	169
	(F)	98.4	132	201.2	110.8	196	81.06	129.6	100.2
	(H)	21	26	39	42.2	36	36	36	27
Patient 6	(W)	72.2	130.2	346.8	226.4	165.6	196	234	222.2
	(F)	51.2	93.3	295	138	120	167	250.2	146
	(H)	36.2	41.3	65.5	37.6	38	34	32	29
Patient 7	(W)	112.6	61.4	310.7	208.3	164.1	103.4	257.5	124.3
	(F)	101.1	45.6	274.4	140.1	60.4	82.9	184.3	97
	(H)	26.1	22.4	76.2	41.1	22.5	24.4	56.2	18
Patient 8	(W)	98.9	118.7	142.1	112.4	141.5	122.6	136.7	117.7
	(F)	91.4	110.6	125.86	109.9	69.4	53.4	127.6	90.1
	(H)	22.5	13.4	29.9	26.2	25.2	26.5	21.7	24.7
Patient 9	(W)	117.9	57.7	293.4	242.6	159.7	82.8	162.9	118.4
	(F)	94.7	41.1	251.3	141.1	63.9	91.6	198.4	148.9
	(H)	29.5	22.4	50.4	29.6	26.2	23.7	29.4	21.4
Patient 10	(W)	59.9	156.4	240	186.3	135.4	161.1	201.8	148.7
	(F)	57.1	123.4	227.6	174.7	132.6	144.4	199.6	141.9
	(H)	19.7	22.9	32.9	33.5	31.5	24.4	27.9	24.4

11.4 FUZZY LOGIC

Fuzzy logic is a multi-valued continuous system providing intermediate values to be defined between conventional evaluations such as true or false, yes or no, and high or low as was used in binary logic. The concept of fuzzy logic has its inception in 1965 by L. A. Zadeh. In fuzzy logic, the available data is represented in the terms of linguistic variables and reasoning, which is easy to apprehend and interpret [18]. The fuzzy logic has extensive application areas and scope of work, due to its suitability to be used as an effective inference system. Such fuzzy inference systems do not only consider two alternatives, instead it takes complete truth space for logical reasoning and propositions. The fuzzy inference system provides middle values to be established between traditional yes or no solutions [19]. In engineering applications, the fuzzy inference system utilizes this incessant subset membership function to transform crimped numerical problems into fuzzy linguistic regions. Fuzzy inference system makes use of conventional linguistics to state variables and fuzzy language procedures to define associations as opposing to employ with numeric variables and mathematical functions. In contrast to the mathematical expression, fuzzy logic offers to use amassed knowledge and experience based on practice rather than theory form. By sustaining the physical inference and effects of all variables, fuzzy logic simulates the complex and nonlinear systems. It is observed that very few works is available in literature which attempts to model pressure at different specific points in terms of patient physical characteristics. Therefore, the present study employed fuzzy logic to model the magnitude of a response variable, such as intra-socket pressure, changes with the values of the independent casual variables, such as weight, height, and SL (for amputee).

The fuzzy membership function characterizes the degree of association between input and output linguistic variables which are represented in Table 11.4. The linguistic terms for different input membership functions are described low (L), medium (M), high (H). Similarly, the output response of pressure is fuzzified in fuzzy sets as very low (VL), low (L), medium (M), high (H), and very high (VH). The membership function and the connected linguistic variables were determined based on the available literature.

TABLE 11.4 Fuzzy Linguistic Term Characteristics

Factors	Linguistic Term	Range
Weight (A)	Very low (VL), low (L), medium (M), high (H), and very high (VH)	52–72(Kg)
Height (B)	Very low (VL), low (L), medium (M), high (H), and very high (VH)	162–175 (cm)
Stump Length (C)	Very low (VL), low (L), medium (M), high (H), and very high (VH)	15–28 (cm)
Pressure at Point 1 (P1)	Very low (VL), low (L), medium (M), high (H), and very high (VH)	21–162 (KPa)
Pressure at Point 2 (P2)	Very low (VL), low (L), medium (M), high (H), and very high (VH)	10–210 (KPa)
Pressure at Point 3 (P3)	Very low (VL), low (L), medium (M), high (H), and very high (VH)	25–360 (KPa)
Pressure at Point 4 (P4)	Very low (VL), low (L), medium (M), high (H), and very high (VH)	25–245 (KPa)
Pressure at Point 5 (P5)	Very low (VL), low (L), medium (M), high (H), and very high (VH)	24–250 (KPa)
Pressure at Point 6 (P6)	Very low (VL), low (L), medium (M), high (H), and very high (VH)	22–212 (KPa)
Pressure at Point 7 (P7)	Very low (VL), low (L), medium (M), high (H), and very high (VH)	20–260 (KPa)
Pressure at Point 8 (P8)	Very low (VL), low (L), medium (M), high (H), and very high (VH)	15–225 (KPa)

11.5 FUZZY MEMBERSHIP FUNCTIONS

The performance of different membership functions for the linguistic terms of input factors and output response was examined based on the least root mean square error (RMSE). From the commonly used membership functions such as triangular, trapezoidal, and Gaussian, this study employs triangular membership function for assessment of input variables and triangular type functions are chosen for output variables as shown in Figure 11.4. The triangular membership function for the output was defined using three parameters m, n, p with vector x:

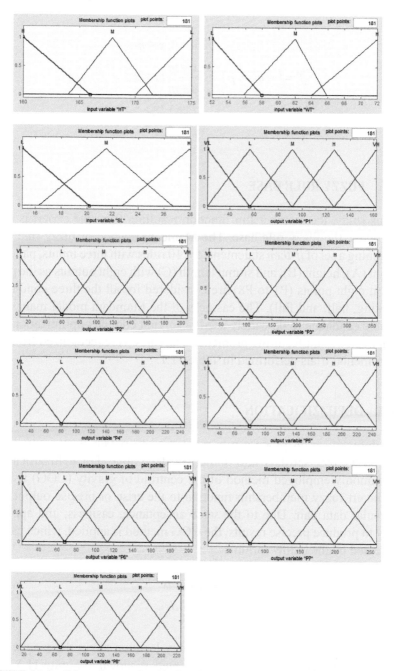

FIGURE 11.4   Triangular membership function for input and output.

$$f = (x; m, n, p) = \begin{cases} 0, x \leq m \\ \dfrac{x - m}{n - m} \, m \leq x \leq n \\ \dfrac{p - x}{p - n} \, n \leq x \leq p \end{cases}$$

11.6 FUZZY RULE BASE

In this study, the membership function resulting from the expert is used to produce the fuzzy rule base. The Mamdani method of fuzzy rule base containing a set of *if-then* statements for 10 rules with three inputs, patient's weight (A), height (B), and stump length (C) with eight outputs as the pressure at eight points (P1 to P8) are considered for all the three conditions (walking, half, and full). The easiness of the Mamdani model makes it a suitable candidate to be widely used in solving complex optimization and complicated manufacturing problems. The rules were recognized from the experimental trials performed in Table 11.1.

11.7 DEFUZZIFICATION

The defuzzification process was performed to obtain fuzzy output and transform fuzzy set data into numerical data value. For defuzzification, a commonly employed method of the centroid of gravity (COG) is used to convert fuzzy membership function to the crisp or precise output of a particular data pair. Due to the wide acceptance, easiness, and applicability to produce precise results, the COG method is applied, which can be expressed by the below formula:

$$x^* = \frac{\int \mu_c(x) x dx}{\int \mu_c(x) dx}$$

where x^* represents the center of gravity location along the x-axis. The fuzzy rule viewer is graphically represented as shown in Figure 11.5. The rows represented individual fuzzy rule and the first three columns depict the input desirability. The last eight columns represent the defuzzified output of pressure at different specific regions.

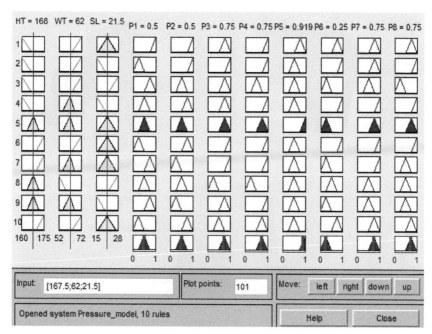

FIGURE 11.5 Graphical representation of fuzzy rule.

11.8 RESULTS AND DISCUSSION

From the experimental results, it is clear that the strongest impact of maximum pressure between stump-socket interfaces is on the PTB (point 3). Therefore, the surface plot predicted by the fuzzy-based model for PTB at point 3 is shown in Figure 11.6. Figure 11.6 depicts the variation of pressure in terms of different physical parameters of an amputee. It can be shown from Figure 11.6 that for 60–65 kg patient weight combining with a higher value of SL is providing maximum pressure of 270 KPa. Similarly, the 60–65 kg weight patients along with the height in the range of 170–175 cm provide a maximum pressure of about 330 KPa. Moreover, the final plot shows that with the height of the patient in the range of 170–175 cm with SL of over 20 cm can produce the maximum pressure of approximately 330 KPa.

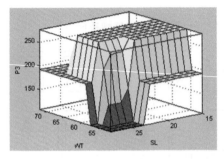
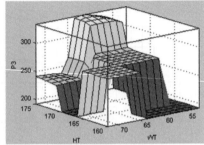
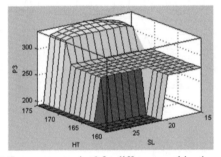

FIGURE 11.6   Predicted pressure at point 3 for different combinations of input parameters.

In the present study, the pressure at different loading conditions is mathematically formulated using the regression analysis technique [20]. Furthermore, regression analysis is carried out on the experimental data collected from the ten patients (Table 11.5) using statistical software Minitab 16. The pressure for three loading conditions was expressed as a function of weight (WT), SL, and height (HT) as shown in Eqs. (11.1)–(11.3).

$$P_{Walking} = 1313.41 - 5.497\ SL - 8.697\ HT + 8.018\ WT \qquad (11.1)$$

$$P_{Full} = 1387.21 - 7.240\ SL - 8.590\ HT + 6.866\ WT \qquad (11.2)$$

$$P_{Half} = 356.359 - 1.384\ SL - 2.404\ HT + 2.047\ WT \qquad (11.3)$$

The analysis of variance (ANOVA) test was executed to statistically investigate the established models for three loading conditions. ANOVA results for the walking condition response model are tabulated in Table 11.5. The ANOVA results for the walking load with model F-value of 20.9 suggest that the model is significant. There is only a 0.01% chance that such a higher model F-value may have happened due to noise. The higher value of the determination coefficient (R^2 = 91.28%) and adjusted determination coefficient (adjusted R^2 = 86.93%) signifies that the model does not clarify only less than 8.72% of the total variation. Hence, this model can be used to navigate the design space.

TABLE 11.5 ANOVA Results for Walking Load Condition

Source	Degree of Freedom	Sum of Squares	Mean Squares	F-Value	P-Value
Regression	3	26416.2	8805.4	20.9	<0.001
SL	1	6321.2	4357.1	10.3	0.0181
HT	1	1776.1	7761.5	18.4	0.0051
WT	1	18318.9	18318.9	43.5	0.0005
Error	6	2522.4	420.4	–	–
Total	9	28938.7	–	–	–
R^2 = 91.28%			Adjusted R^2 = 86.93%		

Table 11.6 shows the ANOVA results for the full load condition response. The ANOVA results for the full load response with a model F-value of 18.1 implying that the model is significant. There is only a 0.01% chance that such a higher model F-value may have occurred due to noise. The higher value of the determination coefficient (R^2 = 90.07%) and adjusted determination coefficient (adjusted R^2 = 85.10%) signifies that the model does not clarify only less than 9.93% of the total variation. Hence, this model can be used to navigate the design space.

ANOVA results for the half load condition response model are listed in Table 11.7. The model F-value of 10.9 with P-value <0.01 implies that the developed model is significant. There is only a 0.01% chance that such a higher model F-value may have occurred due to noise. From Table 11.7, the higher value of the determination coefficient (R^2 = 84.56%) and adjusted determination coefficient (adj. R^2 = 76.84%) signifies that the model does not clarify only less than 15.44% of the total variation. Hence, this model can be used to navigate the design space.

TABLE 11.6   ANOVA Results for Full Load Condition

Source	Degree of Freedom	Sum of Squares	Mean Squares	F-Value	P-Value
Regression	3	25336.8	8445.6	18.1	<0.001
SL	1	9552.7	7559.2	16.2	0.0068
HT	1	2353.5	7573.7	16.2	0.0068
WT	1	13430.6	13430.6	28.8	0.0017
Error	6	2794.4	465.7	–	–
Total	9	28131.1	–	–	–
$R^2 = 90.07\%$			Adjusted $R^2 = 85.10\%$		

TABLE 11.7   ANOVA Results for Half Load Condition

Source	Degree of Freedom	Sum of Squares	Mean Squares	F-Value	P-Value
Regression	3	1752.4	584.1	10.9	<0.01
SL	1	395.2	276.1	5.1	0.0631
HT	1	163.0	593.7	11.1	0.0156
WT	1	1194.2	1194.2	22.3	0.0032
Error	6	319.9	53.3	–	–
Total	9	2072.46	–	–	–
$R^2 = 84.56\%$			Adjusted $R^2 = 76.84\%$		

The accuracy and error percentage determination is a prerequisite for validating newly developed prediction models to know how close the predicted value is from the measured experimental value. In the same context, to check the accuracy and individual error% of the developed fuzzy model in predicting the pressure values is compared with the experimental value at different regions of nine patients (see Table 11.8). The variation of predicted values (E_m) from the actual measured experimental value (E_f) are used to calculate error percentage (E_p) after dividing the absolute difference between the above values by the measured experimental value.

$$E_p = \frac{\left(\left|E_f - E_m\right|\right)}{E_m} X 100\%$$

Furthermore, the accuracy is calculated by determining the closeness of the predicted fuzzy model value to the measured experimental value. The below equation is used, where A_m is the accuracy of the model and n is the total number of considered datasets for finding the average individual accuracy.

$$A_m = \frac{1}{n}\sum_{i=1}^{n}\left(1 - \frac{\left(\left|E_f - E_m\right|\right)}{E_m}\right) X100\%$$

The predicted and measured values of output responses were utilized for the determination of error and accuracy percentage of the applied fuzzy logic model. The highest and lowest error percentage between measured and predicted fuzzy model results for P1–P8 points are 11.57 and 0.11 respectively. Such a consistent and lower error percentage of the overall model specifies that the fuzzy model predicted the outcome which is in close proximity with the experimental results. It is worth noting that the overall model accuracy comes out to be 98.54% in predicting the values of pressure at different points of reconstructed surface. Such high overall model accuracy justifies the utilization of a fuzzy model for successful prediction of output response of pressure at different specific regions in terms of patient physical parameters making it suitable for socket fit.

From the Fuzzy logic-based model for walking load conditions and regression equations fitted for half load, full load, and walking load conditions, it can be said that weight and SL of the patient will affect the pressure developed at the socket interface. Overall, it has perceived that the developed pressure is maximum for the walking condition as compared to full load and half load conditions. It is obvious that while walking condition, the amputee will exert maximum effort and thus the exhibiting maximum pressure at limb/socket interface. The most important outcome of results is that amputee's weight is a most important factor that attributes in determining the amount of pressure at transtibial prosthetic limb/socket interface. The pressure will increase with an increase in amputee's weight. The pressure essential to bear the amputee's weight on a socket would be sovereign of the surface area of the stump/socket interface.

TABLE 11.8 Accuracy and Error Percentage Results

E No.	Output Parameters			Pressure (KPa)							
	Height (cm)	Weight (kg)	Stump Length (cm)	P1				P2			
				Experimental	Fuzzy	Error%	Accuracy %	Experimental	Fuzzy	Error %	Accuracy %
1	172	66	18	161.4	163.10	1.05	98.95	206.5	207.81	0.63	99.37
2	173	70	15	79.4	81.12	2.17	97.83	198.4	199.04	0.32	99.68
3	174	72	28	125.6	125.75	0.12	99.88	111.4	112.67	1.14	98.86
4	171	68	16	108	110.10	1.94	98.06	167.4	171.89	2.68	97.32
5	168	63	23	102	101.72	0.27	99.73	127	125.42	1.24	98.76
6	166	70	23	72.2	74.87	3.70	96.30	130.2	129.65	0.42	99.58
7	162	63	20	112.6	115.05	2.18	97.82	61.4	62.92	2.48	97.52
8	168	52	25	98.9	110.34	11.57	88.43	118.7	119.78	0.91	99.09
9	167	64	19	118	117.15	0.72	99.28	57.7	59.03	2.31	97.69
10	172	66	18	278.5	281.70	1.15	98.85	242	240.82	0.49	99.51
11	173	70	15	315.2	316.12	0.29	99.71	222.5	219.04	1.56	98.44
12	174	72	28	239.5	242.75	1.36	98.64	148.8	152.67	2.60	97.40
13	171	68	16	271.6	274.10	0.92	99.08	229.6	231.89	1.00	99.00

Model accuracy for Pressure = 98.54

TABLE 11.8 (Continued)

E No.	Output Parameters										
	Height (cm)	Weight (kg)	Stump Length (cm)	Pressure (KPa)							
				P1				P2			
				Experimental	Fuzzy	Error%	Accuracy %	Experimental	Fuzzy	Error %	Accuracy %
14	168	63	23	246	244.72	0.52	99.48	176	175.42	0.33	99.67
15	166	70	23	346.8	346.87	0.02	99.98	226.4	229.65	1.44	98.56
16	162	63	20	310.7	311.05	0.11	99.89	208.3	210.92	1.26	98.74
17	168	52	25	142.1	141.34	0.53	99.47	112.4	112.78	0.34	99.66
18	167	64	19	293.4	292.05	0.46	99.54	242.6	241.56	0.43	99.57
19	172	66	18	193.4	191.70	0.88	99.12	165.12	167.81	1.63	98.37
20	173	70	15	146.4	148.12	1.17	98.83	210	209.04	0.46	99.54
21	174	72	28	166.4	165.75	0.39	99.61	79.6	82.67	3.86	96.14
22	171	68	16	136	138.10	1.54	98.46	204	206.89	1.42	98.58
23	168	63	23	243	244.72	0.71	99.29	100	102.42	2.42	97.58
24	166	70	23	165.6	167.87	1.37	98.63	196	195.65	0.18	99.82
25	162	63	20	164.1	169.05	3.02	96.98	103.4	104.92	1.47	98.53
26	168	52	25	141.5	144.34	2.01	97.99	122.6	124.78	1.78	98.22
27	167	64	19	159.7	161.72	1.26	98.74	82.8	83.9	1.33	98.67
28	172	66	18	179.6	181.70	1.17	98.83	179.4	182.8	1.90	98.10

TABLE 11.8 *(Continued)*

E No.	Output Parameters			Pressure (KPa)							
	Height (cm)	Weight (kg)	Stump Length (cm)	P1				P2			
				Experimental	Fuzzy	Error%	Accuracy %	Experimental	Fuzzy	Error %	Accuracy %
29	173	70	15	204.4	206.12	0.84	99.16	210.6	209.04	0.74	99.26
30	174	72	28	153.6	150.75	1.86	98.14	86	88.67	3.10	96.90
31	171	68	16	220.6	218.10	1.13	98.87	224.4	226.89	1.11	98.89
32	168	63	23	184	187.72	2.02	97.98	169	175.42	3.80	96.20
33	166	70	23	234	236.87	1.23	98.77	222.2	224.65	1.10	98.90
34	162	63	20	257.5	261.05	1.38	98.62	124.3	124.92	0.50	99.50
35	168	52	25	136.7	137.34	0.47	99.53	117.7	120.78	2.62	97.38
36	167	64	19	162.9	164.01	0.68	99.32	118.4	119.2	0.68	99.32

Model accuracy for Pressure = 98.54

11.9   CONCLUSIONS

A methodology has been developed using the low-cost piezo-resistive Flexi-Force sensor for quantitatively analyzing the pressure distribution at eight specific regions. Namely, lateral tibia, gastrocnemius, PTB, kick point, medial tibia, medial gastrocnemius, popliteal depression, and lateral gastrocnemius. Ten clinically significant cases have been considered for pressure prediction under different loading conditions. The present study deals with employing fuzzy logic-based modeling for predicting the pressure at specific locations for the limb socket interface for varying physical characteristics of ten clinical cases. The developed fuzzy logic model is validated by performing confirmation experimental trials and an accuracy of 98.53% is obtained in predicting the response of limb socket interface pressure. The present approach significantly evaluates the pressure variation for different loading conditions. The regression analysis has been employed to evaluate the relation been amputee specific physical parameters on maximum pressure condition at the limb-socket interface. From the results, it has found that the amputee's weight and SL are critical parameters and play a major role in pressure prediction. The adopted methodology will help in predicting maximum pressure values at the limb-socket interface using which better designed prosthetic sockets can ensure comfortable socket fitting.

ACKNOWLEDGMENT

The authors thank the patients of below-knee amputees at BMVSS (Jaipur Foot) who voluntarily participated in the study. This study was carried under a project grant (Project Number 180400044) by the Department of Science and Technology, Government of India.

KEYWORDS

- **force-sensing resistors**
- **Fuzzy logic analysis**
- **patella tendon bearing**

- **pressure measurement**
- **socket fitting**
- **stump-socket interface**

REFERENCES

1. Alcaide-Aguirre, R. E., Morgenroth, D. C., & Ferris, D. P., (2013). Motor control and learning with lower-limb myoelectric control in amputees. *J. Rehabil. Res. Dev., 50*(5), 687–698. Pub. Med. PMID: 24013916.

2. Sagar, R., Sahu, A., Sarkar, S., et al., (2016). Psychological effects of amputation: A review of studies from India. *Ind. Psychiatry J., 25*(1), 4–10. doi: 10.4103/0972–6748.196041.

3. Peters, E. J., Lipsky, B. A., Aragón-Sánchez, J., et al., (2016). Interventions in the management of infection in the foot in diabetes: A systematic review. *Diabetes Metab. Res. Rev., 32*, 145–153. doi: 10.1002/dmrr.2706.

4. Al-Fakih, E., Abu, O. N., & Mahmad, A. F., (2016). Techniques for interface stress measurements within prosthetic sockets of transtibial amputees: A review of the past 50 years of research. *Sensors, 16*(7), 1119. doi: 10.3390/s16071119.

5. Sewell, P., Noroozi, S., Vinney, J., et al., (2012). Static and dynamic pressure prediction for prosthetic socket fitting assessment utilizing an inverse problem approach. *Artif. Intell. Med., 54*(1), 29–41. doi: 10.1016/j.artmed.2011.09.005.

6. Carlson, C. E., Mann, R. W., & Harris, W. H., (1974). A Radio telemetry device for monitoring cartilage surface pressures in the human hip. *IEEE Trans Biomed Eng, BME-21*, 257–264. doi: 10.1109/tbme.1974.324311.

7. Al-Fakih, E., Abu, O. N. A., & Mahamd, A. F. R., (2012). The use of fiber bragg grating sensors in biomechanics and rehabilitation applications: The state-of-the-art and ongoing research topics. *Sensors, 12*(12), 12890–12926. doi: 10.3390/s121012890.

8. Williams, R. B., Porter, D., Roberts, V. C., et al., (1992). Triaxial force transducer for investigating stresses at the stump/socket interface. *Med. Biol. Eng. Comput., 30*(1), 89–96. doi: 10.1007/bf02446199.

9. Tiwana, M. I., Shashank, A., Redmond, S. J., et al., (2011). Characterization of a capacitive tactile shear sensor for application in robotic and upper limb prostheses. *Sens. Actuators A Phys., 165*(2), 164–172. doi: 10.1016/j.sna.2010.09.012.

10. Almassri, A. M., Wan, H. W. Z., Ahmad, S. A., et al., (2015). Pressure sensor: State of the art, design, and application for robotic hand. *J. Sens., 1*, 1–12. doi: 10.1155/2015/846487.

11. Luo, Z. P., Berglund, L. J., & An, K. N., (1998). Validation of F-Scan pressure sensor system: A technical note. *J. Rehabil. Res. Dev., 35*(2), 186–191. Pub. Med. PMID: 9651890.

12. Nayak, C., Singh, A., Chaudhary, H., et al., (2016). A novel approach for customized prosthetic socket design. *Biomed. Eng. App. Bas. C, 28*(3). doi: 10.4015/s1016237216500228.

13. Nayak, C., Singh, A., Chaudhary, H., & Unune, D. R., (2017). An investigation on effects of amputee's physiological parameters on maximum pressure developed at the prosthetic socket interface using artificial neural network. *Technology and Health Care, 25*(5), 969–979.

14. Pathak, V. K., Nayak, C., Singh, A. K., & Chaudhary, H., (2016). A virtual reverse engineering methodology for accuracy control of transtibial prosthetic socket. *Biomedical Engineering: Applications, Basis and Communications, 28*(05), 1650037.

15. Pirouzi, G., Abu, O. N. A., Eshraghi, A., et al., (2014). Review of the socket design and interface pressure measurement for transtibial prosthesis. *Sci. World J., 1*, 1–9. doi: 10.1155/2014/849073.

16. Hafner, B. J., & Sanders, J. E., (2014). Considerations for development of sensing and monitoring tools to facilitate treatment and care of persons with lower-limb loss: A review. *J. Rehabil. Res. Dev., 51*(1), 1–14. doi: 10.1682/jrrd.2013.01.0024.

17. Singh, R. K., Gangwar, S., Singh, D. K., & Pathak, V. K., (2019). A novel hybridization of artificial neural network and moth-flame optimization (ANN-MFO) for multi-objective optimization in magnetic abrasive finishing of aluminum 6060. *Journal of the Brazilian Society of Mechanical Sciences and Engineering, 41*(6), 270.

18. Mariajayaprakash, A., Senthilvelan, T., & Gnanadass, R., (2015). Optimization of process parameters through fuzzy logic and genetic algorithm: A case study in a process industry. *Applied Soft Computing, 30*, 94–103.

19. Nunes, I. L., (2010). Handling human-centered systems uncertainty using fuzzy logics-A. *The Ergonomics Open Journal, 3*, 38–48.

20. Neumann, E., Brink, J., Yalamanchili, K., & Lee, J. S., (2013). Regression estimates of pressure on transtibial residual limbs using load cell measurements of the forces and moments occurring at the base of the socket. *J. Ortho and Prosthet., 25*(1), 1–12. doi: 10.1097/JPO.0b013e31827b36.

CHAPTER 12

Experimental Investigation and Optimization of Process Parameters in Oblique Machining Process for Hard-to-Cut Materials Using Coated Inserts

PURNANK BHATT,[1] MIHIR SOLANKI,[1] ANAND JOSHI,[1] and VIJAY CHAUDHARI[2]

[1]*Mechanical Engineering Department, G. H. Patel College of Engineering & Technology, Vallabh Vidyanagar, Gujarat, India*

[2]*Mechanical Engineering Department, CHARUSAT, Changa, Gujarat, India*

ABSTRACT

The high-speed machining process is being widely used in the automobile and aeronautical industries these days where superfine surface finish is the basic requirement. For such a process, a force modeling is done that considers the friction law which is the function of the tool-chip interface temperature which is, in turn, used to calculate different parameters, and ultimately the forces evolved in the machining process. The influence of parameters like velocity and depth of cut on cutting forces is investigated for the empirical relation of the coefficient of friction derived for CRS 1018 for different hard-to-cut materials. For this purpose, tests are carried out turning center for these materials using cryogenically treated cemented carbide tool inserts. The effect of cutting force variation is analyzed experimentally and is compared with the analytical results. Also, optimum cutting force, surface roughness, and power consumption are found out using Taguchi's orthogonal array. For the machining of the Hard-to-cut materials, the temperature aspect plays an important role in the forces evolved during the process. This includes the temperature evolved in the primary shear zone as well as the tool-chip interface.

12.1 INTRODUCTION

Machining of the hard-to-cut materials is a very critical aspect in the area of manufacturing technology. It is for this reason, the prediction of forces becomes the area of concern as the type surface finish obtained is majorly driven by the forces evolved. The prediction of forces may lead us to the prediction of tool wear thereby tool life and the surface finish. When we deal with high-speed machining, the temperature evolved during the process also plays an important in the calculation of the forces. High-speed steel (HSS) is widely used in manufacturing tool materials which are in turn used to machine the workpiece components. So for that purpose, cutting speed of over 120 m/min is considered in our analysis. So, the present study mainly focuses on the effect of different parameters for the oblique cutting process. The analysis of the orthogonal cutting process can be extended to the oblique cutting process using the transformation matrix in Tounsi et al. [1].

To incorporate the thermal effect in the oblique cutting process, some of the parameters like strain rate sensitivity and strain hardening have been considered in Oxley et al. [2]. The primary shear zone has been assumed to have been divided into two unequal parts Astakhov et al. [3]. A coulomb friction law as considered in Moufki et al. [4], has been used which gives the dependence of the coefficient of friction on the tool-chip interface temperature. Based on the empirical relations Moufki et al. [5], a more simplified relation for the coefficient of friction has been considered. For the analysis, the process has been considered as stationary and the analysis is for the one-dimensional flow. In addition to that, the shearing in the band is assumed to be Molinari and Clifton [6]. The chip flow angle η_c is calculated as the function of the cutting velocity V, the rake angle α_n, friction angle λ, and the inclination angle λ_s.

This law considers a constant thickness of the primary shear zone $h = 0.025$ mm. The undeformed chip thickness is small compared to the width of cut and the chip is considered to have formed under approximately plain strain conditions. The chip flow direction is normal to the cutting edge. The chip flow direction is an important factor as it plays an important role in the tool chip contact length. The pressure along the tool chip interface is also considered in the analysis. The anisotropy of the material is not considered in the present analysis as in Johnson and Mellor [7]. The effect of the parameters like the Taylor-Quinney is assumed to be a constant value as the effect of the sliding and sticking zone has been neglected in the present analysis Kushner [8].

The present analysis is done using the MATLAB interface for the mathematical modeling Bhatt et al. [9]. For the mathematical modeling, in order to solve the equations for the stress induction in the primary shear zone, the results seemed to be diverging and the solutions of the equations were obtained in the form of hypergeometric functions. So to simplify the problem, there was an assumption made that the stresses induced at the entrance of the primary shear zone was equal to the yield strength of the material Bhatt, et al. [10]. Taking this value as the reference, the iterations were carried out and the stress variation along the primary shear zone has been obtained.

For the effect of Depth of Cut Cutting and Thrust Force, Dudzinski et al. [18] stated that with an increase in the depth of cut the cutting forces increase for a particular constant value of the velocity. Here the value of velocity is taken as 120 m/min. This trend in analytical modeling has been validated by the experimental results. The numerical values of the experimental results are a bit different from the analytical model as the operating conditions like the feed rate, the diameter of the workpiece, length of the cut, etc. are not specified in Moufki et al. [4] so may differ from the experimental parameters considered. The increase in forces with an increase in the depth of cut can be understood as the force required to deform a larger thickness of the material is large [19]. This is because the number of bonds needed to be broken to initiate plastic deformation are more in the higher depth of cut compared to that in a smaller depth of cut. The range of the thrust force is very negligible as compared to the cutting force. This trend can also be observed in the analytical and experimental results. The study of HSS has been done in this chapter and the comparison for both the materials in terms of experimental results as well as analytical results has been shown.

12.2 ANALYTICAL MODELING OF OBLIQUE CUTTING PROCESSES

The effect of cutting velocity and depth of cut on cutting force, Thrust force, and interface temperature have been studied for Hard-to-cut materials. In the present study, we have considered HSS as our workpiece material. Various coated tool inserts have also been considered in the study and the effect of various cutting parameters on the tool wear have been demonstrated through microscopic images of the cutting edge under consideration.

12.2.1 FOR DRY MACHINING

- Constant Speed, Constant Feed, and Variable Depth of Cut (Figures 12.1–12.3).

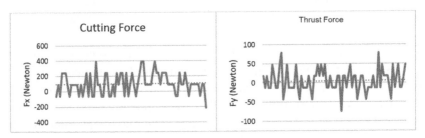

FIGURE 12.1 Cutting force and thrust force plot for conditions: V = 2.638 m/s; DOC = 0.15 mm; feed = 0.06 mm/Rev.

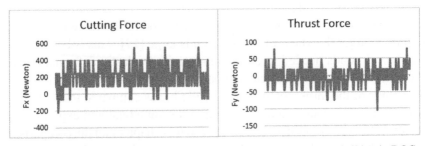

FIGURE 12.2 Cutting force and thrust force plot for conditions: V = 2.638 m/s; DOC = 0.2 mm; feed = 0.06 mm/Rev.

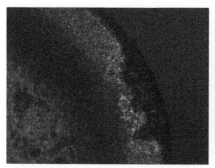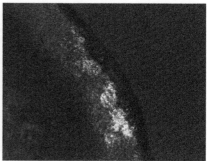

FIGURE 12.3 Cutting force and thrust force plot for conditions: V = 2.638 m/s; DOC = 0.15 mm; feed = 0.06 mm/Rev and V = 2.638 m/s; DOC = 0.2 mm; feed = 0.06 mm/Rev.

Machining parameters play a very important role in the kind of finished component that is manufactured. In the same regards, we have considered a case wherein the cutting speed and the feed are kept constant for the process, and the depth of cut is varied. In such a case, the cutting force obviously increases as shown in the graphs obtained during the machining process. But the variation effect on the thrust force is negligible with an increase in the depth of cut. This means that the surface finish should not have such an effect by variation in the depth of cut during the machining process. This is verified by the surface roughness experimental data shown in Table 12.3.

- Constant Depth of Cut, Constant Feed, and Variable Speed (Figures 12.4–12.6).

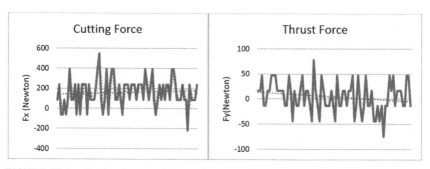

FIGURE 12.4 Cutting force and thrust force plot for conditions: V = 3.72 m/s; DOC = 0.2 mm; feed = 0.02 mm/Rev.

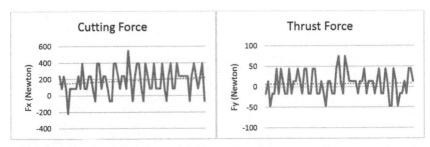

FIGURE 12.5   Cutting force and thrust force plot for conditions: V = 2.638 m/s; DOC = 0.2 mm; feed = 0.02 mm/Rev.

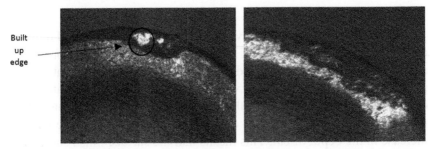

FIGURE 12.6   Cutting force and thrust force plot for conditions: V = 3.72 m/s; DOC = 0.2 mm; feed = 0.02 mm/Rev and V = 2.638 m/s; DOC = 0.2 mm; Feed = 0.02 mm/Rev.

The next case that we have considered is a case wherein the cutting feed and the depth of cut are kept constant for the process and speed is varied. In such a case, the cutting force almost negligibly varying as shown in the graphs obtained during the machining process. The variation effect on the thrust force is negligible as well. But as the length of the cut is increasing, the slight droop can be found in these quantities which maybe because of the thermal softening effect. For a case with higher velocity, a built-up edge formation can be found which suggests that the temperature reached during the machining process was sufficient enough to weld the debris onto the burnt-out part in on the tooltip as shown.

- Constant Speed, Constant Depth of Cut, and Variable Feed (Figures 12.7–12.9).

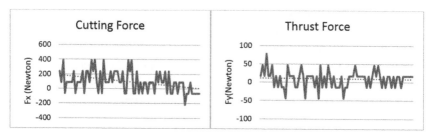

FIGURE 12.7 Cutting force and thrust force plot for conditions: V = 5.2 m/s; DOC = 0.1 mm; feed = 0.04 mm/Rev.

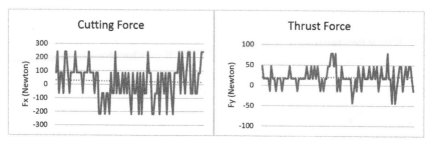

FIGURE 12.8 Cutting force and thrust force plot for conditions: V = 5.2 m/s; DOC = 0.1 mm; feed = 0.02 mm/Rev.

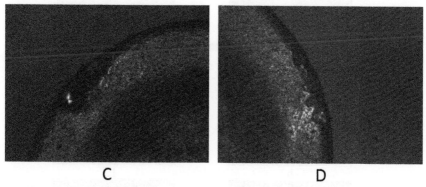

FIGURE 12.9 Cutting force and thrust force plot for conditions: V = 5.2 m/s; DOC = 0.1 mm; feed = 0.04 mm/Rev and V = 5.2 m/s; DOC = 0.1 mm; Feed = 0.02 mm/Rev.

Here cutting speed and the depth of cut are kept constant for the process and feed is varied. In such a case, the cutting force is significantly high for a higher feed rate as shown in the graphs obtained during the machining process. The variation effect on the thrust force is negligible again just like in case of varying depth of cut. But the slight droop can be found in these quantities which maybe because of the thermal softening effect, as the length of cut is increasing. The tool wear in case of higher feed rate is more than that in case of a lower value of feed rate. But again, the wear is not significant if we compare it with other parameters studied in this literature.

12.3   EFFECT ON TOOL WEAR FOR WET AND DRY MACHINING

The above figures clearly suggest the difference in the effect of the tool wear due to wet machining and dry machining. Figures 12.10(a)–(c) are the tool wear profiles observed under the metallurgical microscope for 50x zoom under dry machining conditions. It can be observed that in the case of dry machining, the built-up edge formed can be identified on the burnt-out portion. Built-up edge formation has been shown in one instance under dry machining (Figure 12.10b).

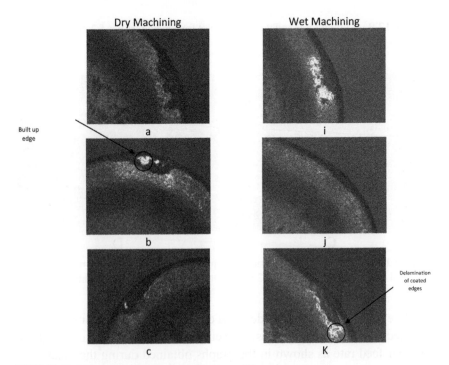

FIGURE 12.10   Tool wear for dry and wet conditions.

We can also observe that there's no or hardly any delamination on the tool tip in the case of dry machining. Similarly, Figures 12.10(i)–(k) are the tool wear profiles observed under the metallurgical microscope for 50x zoom under wet machining conditions. But in the case of wet machining, we can observe the surface delamination of the coating. Though, the figures suggest that there's no built-up edge formed on the tool surface in case of wet machining. This can be explained in a way that for built-up edge to form, proper adhesion of the debris and the tool surface must take place. But because of the use of a coolant, we can't find any built-up edge formed as the coolant takes away the heat generated during the machining process. Thus in absence of heat required to soften the metal and allow it to undergo adhesion, there is no built-up edge formation in wet machining conditions.

12.4 COMPUTATIONAL INVESTIGATION

Figure 12.11 shows the variation in forces (cutting and thrust) and temperature over a primary shear zone. As in line with our experimental results, the force values increase with the cutting velocity. But an important observation that one can go for from the force plots is that at a lower value of cutting velocity, the length of the tool chip interface is lower. Also, the temperature at the interface is higher for higher velocity plots as observed in the figure. As the velocity increases, the point of maximum temperature shifts farther from the tool tip. This aspect is yet to be analyzed in the experimental results which can be considered as the future work for this course which may help us the probable location of the built-up edge formation.

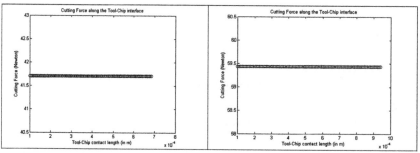

FIGURE 12.11   (A, B) Cutting force, thrust force, and temperature variations for two different conditions of cutting speed.

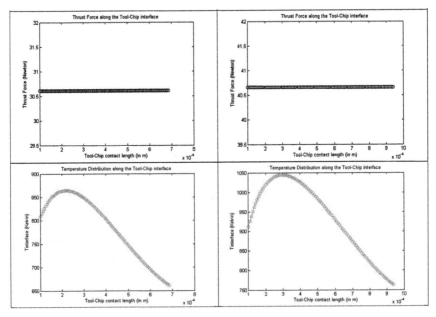

FIGURE 12.11 (C, D, E, F) Cutting force, thrust force, and temperature variations for two different conditions of cutting speed.

12.5 EXPERIMENTAL INVESTIGATION

The design of experiment study is carried out to plan and construct the machining of hard-to-cut material. For this study, Taguchi's L-18 orthogonal array is selected with five variables and mixed-level design. Taguchi's orthogonal array gives near the same result as full factorial design, Solanki and Desai [11]. Minitab statistical software is used to construct and analyses the experiment. Parameters selected for this experiment are cutting conditions, cutting tool, speed (rpm), feed rate (mm/rev.), and depth of cut (mm). For high speed, machining it is important to do the machining above the cutting velocity of 120 m/minute. The diameter of the workpiece is 71 mm and corresponding to that minimum chuck speed is 538 rpm. The outlet of design from the Minitab is as shown in Table 12.1. Cutting conditions are selected as wet and dry; cutting tools used for this experiment are-CNMG235, CNMG 4235 and 710; feed rate based on initial study selected as 0.02, 0.04 and 0.06 m/revolution; depth of cut is selected as 0.10, 0.15, and 0.20 mm; cutting speed is selected as 710, 1000, and 1400 rpm.

TABLE 12.1 Outlet of Design of Experiment (L-18 OA)

Sl. No.	Working Conditions	Tool Tip	Speed (rpm)	Feed Rate (mm/rev)	Depth of Cut (mm)
1.	Wet	CNMG-710	710	0.02	0.1
2.	Wet	CNMG-710	1000	0.04	0.15
3.	Wet	CNMG-710	1400	0.06	0.2
4.	Wet	CNMG-235	710	0.02	0.15
5.	Wet	CNMG-235	1000	0.04	0.2
6.	Wet	CNMG-235	1400	0.06	0.1
7.	Wet	CNMG-4325	710	0.04	0.1
8.	Wet	CNMG-4325	1000	0.06	0.15
9.	Wet	CNMG-4325	1400	0.02	0.2
10.	Dry	CNMG-710	710	0.06	0.2
11.	Dry	CNMG-710	1000	0.02	0.1
12.	Dry	CNMG-710	1400	0.04	0.15
13.	Dry	CNMG-235	710	0.04	0.2
14.	Dry	CNMG-235	1000	0.06	0.1
15.	Dry	CNMG-235	1400	0.02	0.15
16.	Dry	CNMG-4325	710	0.06	0.15
17.	Dry	CNMG-4325	1000	0.02	0.2
18.	Dry	CNMG-4325	1400	0.04	0.1

With process parameters mentioned in Table 12.1, the response variable selected for optimization study are surface roughness and material removal rate.

12.6 OPTIMIZATION OF SURFACE ROUGHNESS AND MATERIAL REMOVAL RATE

With the input process parameters mentioned in Table 12.1, experiments are carried out in random order. The machining process is carried out on a conventional lathe machine. Surface roughness is measured with the help of sophisticated surface texture measuring instrument SV-2100 from Mitutoyo Corporation. Measurement conditions are selected as sample length of 25 mm, the total number of 8 samples, and pitch of 1.0 μm with filter setting of Gaussian element and speed of stylus movement is

0.5 mm/second. The outcome profile of the surface roughness for all the iteration is shown in Table 12.2. To measure the material removal rate during the turning process, a cut for a specific length of 25 mm is carried out. Weight of specimens before and after the turning process is measured. The outcome of the experiment results is mentioned in Table 12.3.

TABLE 12.2 Surface Texture Profile for L-18 OA

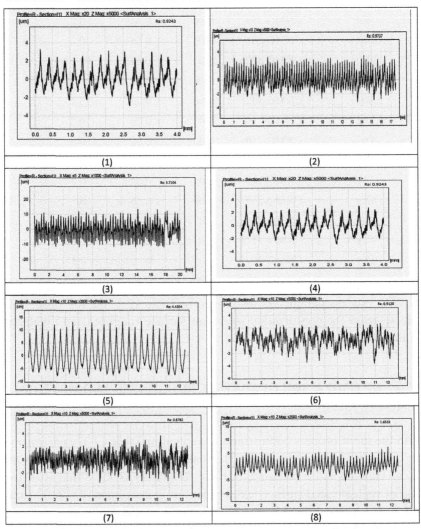

TABLE 12.2 (*Continued*)

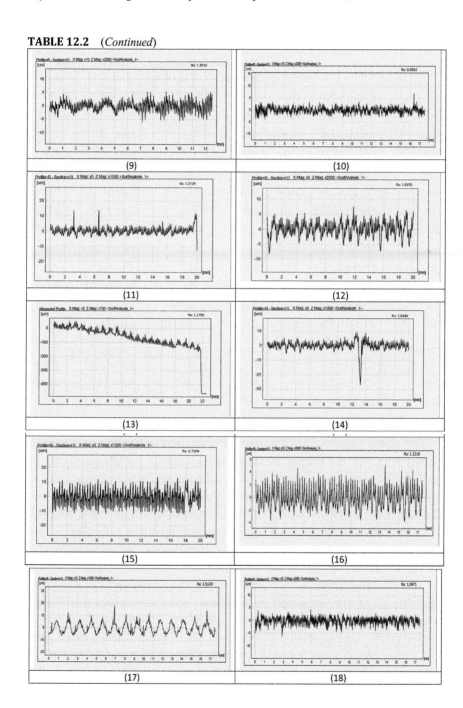

TABLE 12.3 Outcome Results of MRR and Ra for L-18 OA

Conditions	Tool	Speed (rpm)	Feed Rate (m/rev)	Depth of Cut (mm)	MRR (mm³/sec)	Ra (μm)
Wet	CNMG-710	710	0.02	0.1	9.8×10^{-7}	0.9234
Wet	CNMG-710	1000	0.04	0.15	2.45×10^{-7}	0.9737
Wet	CNMG-710	1400	0.06	0.2	3.68×10^{-7}	3.7104
Wet	CNMG-235	710	0.02	0.15	9.8×10^{-7}	0.9234
Wet	CNMG-235	1000	0.04	0.2	1.229×10^{-7}	4.4354
Wet	CNMG-235	1400	0.06	0.1	1.228×10^{-7}	0.912
Wet	CNMG-4325	710	0.04	0.1	3.68×10^{-7}	0.8782
Wet	CNMG-4325	1000	0.06	0.15	3.68×10^{-7}	1.6535
Wet	CNMG-4325	1400	0.02	0.2	24.5×10^{-7}	1.3932
Dry	CNMG-710	710	0.06	0.2	49×10^{-7}	0.9363
Dry	CNMG-710	1000	0.02	0.1	11.06×10^{-7}	1.5729
Dry	CNMG-710	1400	0.04	0.15	8.6×10^{-7}	1.8205
Dry	CNMG-235	710	0.04	0.2	54×10^{-7}	1.175
Dry	CNMG-235	1000	0.06	0.1	44.2×10^{-7}	1.8346
Dry	CNMG-235	1400	0.02	0.15	3.68×10^{-7}	3.7104
Dry	CNMG-4325	710	0.06	0.15	4.9×10^{-7}	1.221
Dry	CNMG-4325	1000	0.02	0.2	2.45×10^{-7}	3.563
Dry	CNMG-4325	1400	0.04	0.1	6.14×10^{-7}	1.0971

The significance of the parameters is studied on the basis of experimental data. As the output characteristic (surface roughness) is to be studied for this process, Taguchi's signal to noise(S/N) ratio concept is utilized. Because surface roughness is a small-the-best type of characteristic, the S/N ratio formula used for this analysis is $10\log(Y/S^2)$, where y is considered as average and s is the standard deviation [16]. The values of S/N ratio are calculated for all 18 experiments and these values of S/N ratio further analyzed for checking the statistical significance of parameters. The graph of S/N ratio for surface roughness is shown in Figure 12.12. From the graph of S/N ratio, optimum values for process parameters should be selected at a higher level. Working condition at level-1 (wet), tool at level-1 (CNMG-7115), feed at level-1 (0.20 mm/rev), the speed at level-3 (1400 rpm), and depth of cut at level-1 (0.10 mm), should be selected to minimize the roughness of the surface, Wu and Hamada [12].

Main Effects Plot for SN ratios
Data Means

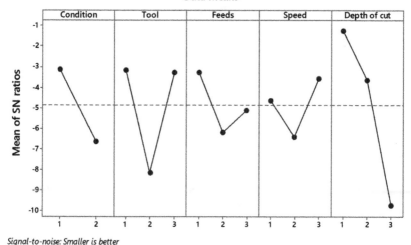

Signal-to-noise: Smaller is better

FIGURE 12.12 S/N ratio plot for surface roughness (Ra).

As the material removal rate is a larger-the-best type of characteristic, the S/N ratio formula used is $10\log(1/Y)/S^2$. The values of S/N ratio are calculated for all experiments. The plot of S/N ratio for MRR is shown in Figure 12.13. From the graph of S/N ratio, optimum values for process parameters should be selected at a higher level. Working condition at level-2 (dry), cutting tooltip at level-2 (CNMG-711), cutting speed at level-1 (710 rpm), Feed rate at level-1 (0.02 mm/rev), and depth of cut at level-3 (0.20) should be selected to maximize MRR.

To investigate the process parameters, ANOVA (analysis of variance) are performed, Montgomery and Runger [13], as shown in Table 12.4. It is not convenient to identify the significance level of process parameters based on S/N ratio graph. Thus, to establish statistical significance, the confidence interval is assessed with 95% ($\alpha = 0.05$) for the differences in mean. Using Minitab software depth of cut had DF (degree of freedom) (2), the adjacent sum of square (Adj SS) (26.985), adjacent mean square (Adj MS) (26.9853), F-value (4.63) and P-value (0.042). Since $p \leq 0.05$, depth of cut is a significant factor for surface roughness, Gijo and Scaria [14]. For working conditions, cutting tooltip, speed, and feed rate have $p \geq 0.05$, they are not statistically significant.

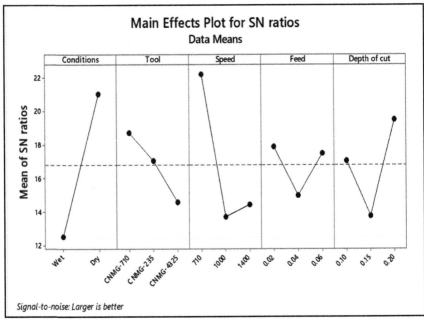

FIGURE 12.13 S/N ratio plot for MRR.

TABLE 12.4 Analysis of Variance for Surface Roughness

Source	DF	Adj SS	Adj MS	F-Value	P-Value
Condition	1	6.879	6.8791	1.21	0.364
Tool tip	1	0.001	0.0014	0.00	0.298
Feed rate	1	0.971	0.9711	0.17	0.690
Speed	1	0.276	0.2756	0.05	0.831
Depth of cut	1	26.985	26.9853	4.64	0.042
Error	12	69.871	5.8226	–	–
Total	17	69.871	–	–	–

To assess the significance of the process parameter on MRR, ANOVA is conducted as shown in Table 12.5. The ANOVA study is conducted with 95% of the confidence interval. As the p-value for the parameters has greater than 0.05, the null hypothesis gets accepted. Hence, none of the factors is significant in this study for MRR.

TABLE 12.5 Analysis of Variance for MRR

Source	DF	Adj SS	Adj MS	F-Value	P-Value
Condition	1	854.0	854.0	3.35	0.092
Tool tip	1	128.3	128.3	0.50	0.492
Feed rate	1	579.0	579.0	2.27	0.157
Speed	1	171.7	171.7	0.67	0.428
Depth of cut	1	287.6	287.6	1.13	0.309
Error	12	3059.9	255.0	–	–
Total	17	3059.8	–	–	–

KEYWORDS

- **analysis of variance**
- **degree of freedom**
- **dry machining**
- **high-speed steel**
- **machining parameters**
- **thrust force**

REFERENCES

1. Tounsi, N., Vincenti, J., Otho, A., & Elbestawi, M. A., (2002). From the basics of orthogonal metal cutting towards the identification of the constitutive equation. *Int. J. Mach. Tools Manuf., 42*(2), 1373–1383.
2. Oxley, P. L. B., & Welsh, M. J. M., (1963). Calculating the shear angle in orthogonal metal cutting from fundamental Stress, strain, and strain rate properties of the work material. In: *Proceedings of the 4th International Machine Tool Design and Research Conference* (pp. 73–86). Oxford.
3. Astakhov, V. P., Osman, M. O. M., & Hayajneh, M. T., (2001). Re-evaluation of the basic mechanics of orthogonal metal cutting: Velocity diagram, virtual work equation, and upper bound theorem. *Int. J. Mach. Tools Manuf., 41*, 393–418.
4. Moufki, Molinari, A., & Dudzinski, D., (1998). Modeling of orthogonal cutting with a temperature-dependent friction law. *J. Mech. Phys. Solids, 46*(10) 2103–2138.
5. Moufki, A., Dudzinski, D., Molinari, A., & Rausch, M., (1999). Thermo viscoplastic modeling of oblique cutting: Forces and chip flow predictions. *International Journal of Mechanical Sciences, 42*(2000), 1205–1232.

6. Molinari, A., & Clifton, R. J., (1987). Analytical determination of shear localization in thermoviscoplastic materials. *Journal of Applied Mechanics, 54,* 80612.

7. Johnson, W., & Mellor, P. B., (1983). *Engineering Plasticity.* John Wiley & Sons, New York.

8. Kushner, V. S., (1982). *Thermo Mechanical Approach in Metal Cutting.* Irkutsk Univ. Pabl. Ltd, Irkutsk, Russia.

9. Bhatt, P., Tewari, A., & Raval, H. K., (2018). Modeling of oblique cutting process using coulombs friction law for Ti-6Al-4V Alloy. *Materials Today: Proceedings, 5*(11), 23596–23602.

10. Bhatt, P., Tewari, A., & Raval, H. K., (2018). Parametric study and modeling of orthogonal cutting process for AISI 4340 and Ti-6Al-4V alloy. *Materials Today: Proceedings, 5*(2), 4730–4735.

11. Solanki, M., & Desai, D., (2015). Comparative study of TQM and six sigma. *International Journal of Industrial Engineering and Technology, Trans Stellar Publication, 5*(4), 1–6.

12. Wu, C. F. J., & Hamada, M., (2011). *Experiments-Planning, Analysis, and Parameter Design Optimization.* John Wiley, New York, NY.

13. Montgomery, D. C., & Runger, G. C., (2007). *Applied Statistics and Probability for Engineers* (4th edn.). John Wiley & Sons, Inc, London.

14. Gijo, E. V., & Scaria, J., (2010). Reducing rejection and rework by application of six sigma methodology in the manufacturing process. *International Journal of Six Sigma and Competitive Advantage, 6*(1/2), 77–90.

15. Binglin, L., Yujin, H., Xuelin, W., Chenggang, L., & Xingxing, L., (2011). An analytical model of oblique cutting with application to end milling. *Machining Science and Technology: An International Journal, 15*(4), 453–484. doi: 10.1080/10910344.2011.620920.

16. Taguchi, G., (1987). *Systems of Experimental Design* (Vols. 1 and 2). UNIPUB and American Supplier Institute, New York, NY.

17. Astakhov, V. P., & Osman, M. O. M., (1996). An analytical evaluation of the cutting forces in self-piloting drilling using the model of the shear zone with parallel boundaries. Part 1: Theory. *International Journal of Machine Tools and Manufacture, 36*(11), 1187–1200.

18. Dudzinski, D., & Molinari, A. (1997). A modelling of cutting for viscoplastic materials. *Int. J. Mech. Sci., 39*(4), 369–389.

19. Merchant, M. E. (1944). Basic mechanics of the metal cutting process. *ASME J. Appl. Mech. 66,* 168–175.

CHAPTER 13

Mechanical Design of a Slider-Crank Mechanism for a Knee Orthotic Device Using the Jaya Algorithm

RAMANPREET SINGH and VIMAL KUMAR PATHAK

Department of Mechanical Engineering, Manipal University Jaipur, Jaipur, Rajasthan, India,
E-mail: ramanpreet.singh@jaipur.manipal.edu (R. Singh)

13.1 INTRODUCTION

Human gait is a complex function that requires repetitive and coordinated rhythmic movement of lower-limbs. Its control is dependent on the coordination of the musculoskeletal system, locomotion mechanism, and motor control. Any disturbance in their coordination may lead to stroke. Stroke destroys many the cortical neurons whereas remaining neurons are temporarily affected. The temporarily affected neurons may regain functioning to some extent. Therefore, post-stroke patients are suggested to undergo rehabilitation and try to relearn walking. Typically, this is done with the help of manual physiotherapies and with assistive devices [1, 2]. Traditional therapies are done manually with the help of therapists, treadmills, bodyweight support systems, etc. However, they are physically demanding for therapists. Also, frequent and consistent training sessions may not be available which may leave a patient with permanent disability [3].

Various treadmill-based exoskeletons have been explored over the years. Typically, these are multi-degree of freedom (DF) devices that contain mechanisms, actuators, and control technology for manipulating users' lower limb motion while walking. Some of the treadmill-based rehabilitation devices are discussed here. The ReoAmbulatorTM which is commercialized by Motorika USA Inc. powered to lift a patient from a wheelchair and transports the patient over the treadmill [4]. Lokomat is developed from the prototype of driven gait orthosis and it is provided with virtual reality environment along with audio and visual biofeedback which is commercially available [5] while driven gait orthosis [6], LokoHelp [7], Alex [8], and Lopes [9] are among other treadmill-based rehabilitation devices. Also, the treadmill-based exoskeletons are bulky and they are often used in rehabilitation centers and hospitals. The other categories of exoskeletons may be over-ground and portable exoskeletons. The mobile base over-ground rehabilitation exoskeletons may consist of a mobile base, a BWS, and joint level assistance to provide comfort to the patients for rehabilitation. They do not restrict training to the treadmill or a confined area rather they allow patients to regain their natural walk. In addition, the patients move voluntarily despite considering a predetermined pattern for moving. Some of the over-ground exoskeletons are explored to identify the mechanism used for the joints. EXPOS developed by Sogang University is used especially for the elderly and the patients [10]. Another version of the EXPOS which is known as SUBAR (Sogang University biomedical assistive robot) may also be used for over-ground rehabilitation [11]. LEER (lower extremity exoskeleton robot) [12], NatTUre-gaits [13], WalkTrainer™ [14], and Kine Assist robotic device [15] are among other rehabilitation devices which may also be used.

Another category of lower-limb robotic devices is portable rehabilitation exoskeletons or assistive devices. These exoskeletons are mobile and do not require any base or treadmill. As opposed to treadmill-based exoskeletons, they are lightweight and easy to don and doff. Their simple and small structure makes them relatively more comfortable in comparison with treadmill-based and mobile-based overground exoskeletons. Besides, one of the most notable features of the portable exoskeletons is that they allow natural walking, and the power source is attached to the exoskeleton for actuating the joints. In addition, the users require crutches along with the exoskeleton during walking, because of their impaired physical ability. Some of these portable multi-DOF exoskeletons are explored here. ReWalk [16] and HAL [17] are the commercially available portable exoskeletons that can be used with crutches for rehabilitation. Other portable rehabilitation devices developed by researchers include Powered Orthosis of Vanderbilt University

[18], modified motor-powered gait orthosis [19], powered gait orthosis (PGO) [20], knee-ankle-foot robot [3], etc.

It is found that linkage mechanisms play a vital role in the actuation of mechanism, gait speed, step length, etc. Hence, it is also worthwhile to investigate the area of assisting devices or orthosis for knee joint while synthesizing a mechanism for supporting an injured knee, or gait rehabilitation. Typically, it is used to correct the functions of physically impaired patients. Some of the important features of an orthosis are restraining joint mobility, correcting limb malformations, assistance in ambulation, stability, etc. [21]. Various orthoses that are used for assisting an injured knee or rehabilitating gait are explored. The orthosis/ knee brace uses a three-point fixation system to avoid hyperextension and effectively control hyperextension [22]. Another device is an off-the-shelf knee brace with a hinged between the thigh and shank which can be used by osteoarthritis patients [23]. Another orthosis that can be used for providing relief to osteoarthritis patients can be an adjustable unloader knee brace which uses a polycentric joint between thigh and leg segments. In addition, this novel knee brace does not require straps for providing the needed moment [24].

Besides, the orthoses can be extended to ankle and foot and those types of orthoses are called knee-ankle-foot orthoses (KAFO). KAFO contains a cam mechanism with friction rings and lock that enable the KAFO to lock the knee joint at any position to assist the patients with knee flexion contractures [25]. Another KAFO uses a four-bar linkage for coupling the knee and ankle movement [26]. In addition, actuators with linkage mechanism at the knee joints are also found [27–29]. A four-bar linkage actuator for knee assisting devices can be used to mimic the motion of the human knee joint for the rehabilitation of hemiplegic patients [30]. Devices which couple the knee and ankle movement through linkage mechanisms and which use linkage mechanisms for actuation are among others. Thus, the mechanisms play a vital role in the functioning of the exoskeleton and assisting devices. A mechanism can be employed in rehabilitation devices, bipeds, exoskeletons, etc.

Single DF walking mechanism is another area in which researchers are working actively. Various synthesis techniques and mechanisms have been explored for designing them. A cam-driven mechanism in which a cam system attached to the body frame connects the feet of the robot through a pantograph mechanism that can be used [31]. A six-bar linkage can be used to approximate the femur and tibia motions while a third leg can be included to ensure the frontal stability [32]. Alternatively, a six-bar Stephenson III mechanism can be used for designing a walking linkage. The mechanism can be synthesized in two stages; four-bar linkage that generates the inverted gait should be synthesized in the first stage followed by the synthesis of a dyad that inverts and magnify the gait [33]. Another six-bar linkage, i.e., Klann linkage can also be used for producing a variety of gaits [34].

It is found that most of the devices are treadmill-based gait rehabilitation systems, overground, and portables devices that have multi-degrees of freedom, in addition, they use a single-axis, revolute joint at the knee, which allows only rotational motion. The devices that use linkage mechanisms at the knee, hip, ankle, and foot are rare and they require the large value of peak torque to their actuation design. Therefore, in this chapter, an optimized design of the slider-crank knee orthotic device is proposed. Its mechanical design is inspired from the design of a compact portable knee-ankle-foot robot which can be referred in detail from Ref. [3].

The remainder of the chapter is structured as follows. Section 13.2 discusses the biomechanics of the human knee joint. Section 13.3 illustrates the mechanism design and optimization problem formulation. The optimization algorithm required to solve the optimization problem is presented in Section 13.4. Lastly, conclusions are outlined in Section 13.5.

13.2 BIOMECHANICS OF HUMAN KNEE

Comprehension of physiological gait patterns is crucial for defining the trajectories to guide walking. Human walking (gait) pattern varies with individuals; hence, it should be studied for many subjects. The kinematics and kinetics data for human gait can be obtained from the normative gait database [35, 36]. Typically, it is presented in a statistical average and standard deviation of several sets of values normalized for a complete gait cycle. The measurement techniques and subjects may differ for distinct research groups; therefore, the reported data may differ. Figures 13.1 and 13.2 demonstrate the biomechanics of the human lower limb in terms of knee flexion/extensions and torques.

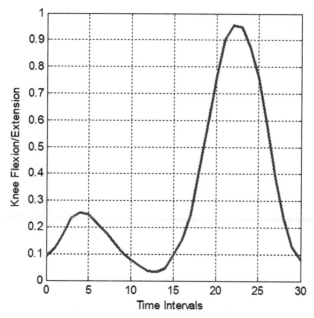

FIGURE 13.1 Average knee flexion/extension.

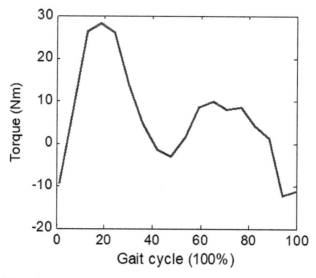

FIGURE 13.2 Joint torque.

13.3   MECHANISM DESIGN AND OPTIMIZATION PROBLEM FORMULATION

A slider-crank mechanism is used in this design for the knee joint. The Schematic diagram of the slider-crank mechanism for the orthotic knee joint is shown in Figure 13.4. An optimization problem is formulated to minimize the required peak force by the actuator. To obtain an adequate and safe motion range, the physical structure and geometrical constraints are adapted from Ref. [3]. Figures 13.3 and 13.4 show various definitions of the slider-crank mechanism for the knee joint. is the crank angle when the relative angle between femur and tibia, $\theta_{knee,}$ becomes 0. Also, to ensure that at the peak value of the required torque the crank and connecting rod becomes perpendicular, consequently, it maximizes the length of the lever-arm to apply force a constraint is used as follows:

$$\theta_{kneemax} + \alpha_k = 90° \qquad (13.1)$$

In which, $\theta_{kneemax}$ represents knee joint orientation corresponding to maximum torque.

To optimize the geometrical parameters of the four-bar slider-crank mechanism, and to ensure the allowable range of motion, following kinematics of the slider-crank mechanism is considered:

$$d \cos \theta_{k1} = c_1 + c_2 \cos \theta_{k2} \qquad (13.2)$$

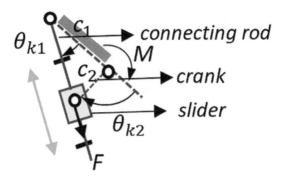

FIGURE 13.3   Various definitions of slider-crank mechanism.

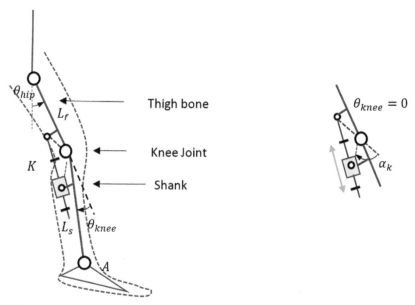

FIGURE 13.4   Schematic of the orthotic knee joint with the slider-crank mechanism.

$$d \sin \theta_{k1} = c_2 \sin \theta_{k2} \qquad (13.3)$$

where, $\theta_{k2} = a \cos \cos \left(\dfrac{d_{min}^2 - c_1^2}{2c_1 c_2} \right)$; also $\theta_{k2} = \alpha_k + \theta_{knee}$

The torque generated at the knee joint can be posed as:

$$M = Fc_2 \sin (\theta_{k2} - \theta_{k1}) \qquad (13.4)$$

Since the only unknown in Eq. (13.4) is . Therefore, it is posed as minimization of peak actuator force problem in the form of an infinite norm, as follows:

$$\text{Min } f(x) = F_{\infty} = \left\| \frac{M}{c_2 \sin \sin (\theta_{k2} - \theta_{k1})_{\infty}} \right\|$$

The geometrical constraints for the above objective function can be posed as follows, as in Ref. [3]:

Subject to $d_{max} < c_1 + c_2$

$$d_{min} > c_1 - c_2$$

$$c_1 - c_2$$

$$\theta_{kneemax\_req} \leq \theta_{k2} - \alpha_k \leq \theta_{kneemax\_limit}$$

$$\theta_{kneemin\_limit} \leq \theta_{k2} - \alpha_k \leq \theta_{kneemin\_req}$$

13.4 OPTIMIZATION ALGORITHM AND RESULTS

In this section, an optimization algorithm, Jaya, is used to solve the problem formulated in the previous section (as adopted from Ref. [3]). Jaya [37] is a simple algorithm that tries to move the candidate solution towards the best candidate and away from the worst candidate.

The algorithms begin with the random population initialization using Eq. (135) in which x^t_{ji}, is the jth candidate for $j = 1, ..., c$, that contains design variables ($i = 1, ... d$). L_i and U_i are the lower and upper limits on the ith design variable and t represents t^{th} iteration.

$$x^t_{ji} = L_i + rand\left(U_i - L_i\right) \tag{13.5}$$

From the population of candidates, best, and worst candidates are identified. These best and worst candidates are utilized to move the candidate solution towards the best and away from the worst solution. This can be achieved by using Eq. (13.6) as follows:

$$x^{t+1}_{ji} = x^t_{ji} + rand_1\left(x_{best} - \left|x^t_{ji}\right|\right) - rand_2\left(x_{best} - \left|x^t_{ji}\right|\right) \tag{13.6}$$

In which, $rand_1$ and $rand_2$ are the random numbers between [0, 1].

The updated population of the candidates is then compared with the previous population of candidates. A new population is formed by the greedy selection of candidates between two populations. Lastly, the termination criterion is checked and this step completes the first iteration of Jaya algorithm. The details of the Jaya algorithm are provided in Figure 13.5.

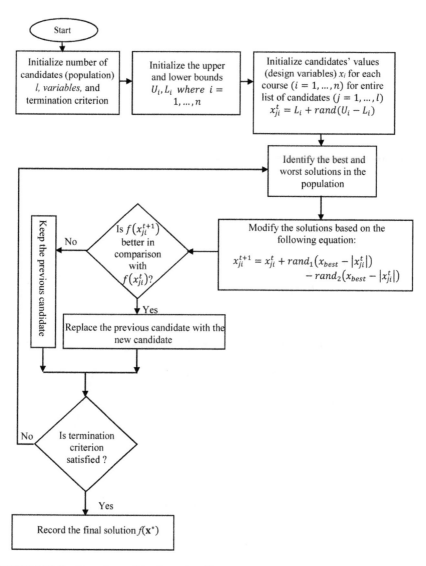

FIGURE 13.5 Flowchart of the Jaya algorithm.

The algorithm is implemented to minimize peak force for the design of a slider-crank mechanism. The algorithm helped in reducing the required actuator force. The results of the new design after optimization are presented in Table 13.1.

TABLE 13.1   Optimized Parameters of Robotic Knee Joint

Parameters	Knee Joint [3]	Knee Joint Proposed
Initial Angle α_k (*degrees*)	70	73.17
$\theta_{kneemix\_req} - \theta_{kneemax\_req}$ (limiting range)	0–120	8.67–74
c_1 (mm)	168	177.6
c_2 (mm)	45	53
Peak Force (N)	687	554.176

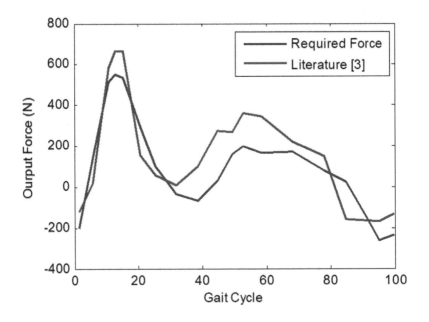

FIGURE 13.6   Required output force.

Figure 13.6 shows the comparison between the required output forces by the slider-crank mechanism with that of the mechanism presented in Ref. [3]. It is found that there is a reduction of 19% in the required peak force using the proposed slider-crank mechanism in the orthotic device.

13.5 CONCLUSIONS

The chapter proposed a slider-crank linkage for an orthotic device that required less peak force for actuation. The mechanism is synthesized using the optimal synthesis technique. An optimization inspired problem and geometrical constraints are formulated based on the notion presented in the literature. A metaphor-less algorithm, Jaya, is used to solve the optimization problem. It is found that the required peak force has reduced to around 19% of the peak force presented in the literature. The proposed slider-crank mechanism can be used for the orthotic devices to provide support to the injured knee joint.

KEYWORDS

- **biomechanics**
- **knee-ankle-foot orthoses**
- **lower extremity exoskeleton robot**
- **optimization algorithm**
- **powered gait orthosis**
- **Sogang University biomedical assistive robot**

REFERENCES

1. Kora, K., Stinear, J., & McDaid, A., (2017). Design, analysis, and optimization of an acute stroke gait rehabilitation device. *Journal of Medical Devices, 11*(1), 014503.
2. Langhorne, P., Stott, D., Knight, A., Bernhardt, J., Barer, D., & Watkins, C., (2010). Very early rehabilitation or intensive telemetry after stroke: A pilot randomized trial. *Cerebrovascular Diseases, 29*(4), 352–360.
3. Chen, G., Qi, P., Guo, Z., & Yu, H., (2016). Mechanical design and evaluation of a compact portable knee-ankle-foot robot for gait rehabilitation. *Mechanism and Machine Theory, 103*, 51–64.

4. West, R. G., (2004). Inventor; Health South Corporation, assignee. *Powered Gait Orthosis and Method of Utilizing Same.* United States patent US 6,689,075.

5. Riener, R., Lunenburger, L., Jezernik, S., Anderschitz, M., Colombo, G., & Dietz, V., (2005). Patient-cooperative strategies for robot-aided treadmill training: First experimental results. *IEEE Transactions on Neural Systems and Rehabilitation Engineering, 13*(3), 380–394.

6. Colombo, G., Joerg, M., Reinhard, S. B. N., & Dietz, V., (2000). Treadmill training of paraplegic patients using a robotic orthosis. *Development, 37*(6), 693–700.

7. Freivogel, S., Mehrholz, J., Husak-Sotomayor, T., & Schmalohr, D., (2008). Gait training with the newly developed 'Loko Help'-system is feasible for non-ambulatory patients after stroke, spinal cord, and brain injury: A feasibility study. *Brain Injury, 22*(7–8), 625–632.

8. Banala, S. K., Agrawal, S. K., & Scholz, J. P., (2007). Active leg exoskeleton (ALEX) for gait rehabilitation of motor-impaired patients. In: *Rehabilitation Robotics* (pp. 401–407). ICORR 2007, IEEE 10th International Conference on 2007 Jun 13 IEEE.

9. Veneman, J. F., Kruidhof, R., Hekman, E. E., Ekkelenkamp, R., Van, A. E. H., & Van, D. K. H., (2007). Design and evaluation of the LOPES exoskeleton robot for interactive gait rehabilitation. *IEEE Transactions on Neural Systems and Rehabilitation Engineering, 15*(3), 379–386.

10. Kong, K., & Jeon, D., (2006). Design and control of an exoskeleton for the elderly and patients. *IEEE/ASME Transactions on Mechatronics, 11*(4), 428–432.

11. Kong, K., Moon, H., Hwang, B., Jeon, D., & Tomizuka, M., (2009). Impedance compensation of SUBAR for back-drivable force-mode actuation. *IEEE Transactions on Robotics, 25*(3), 512–521.

12. Guo, Z., Fan, Y., Zhang, J., Yu, H., & Yin, Y., (2012). A new 4M model-based human-machine interface for lower extremity exoskeleton robot. In: *International Conference on Intelligent Robotics and Applications* (pp. 545–554). Springer, Berlin, Heidelberg.

13. Wang, P., Low, K. H., & Tow, A., (2011). Synchronized walking coordination for impact-less footpad contact of an overground gait rehabilitation system: NaTUre-gaits. In: *Rehabilitation Robotics (ICORR)* (pp. 1–6). IEEE International Conference on 2011 Jun 29. IEEE.

14. Bouri, M., Stauffer, Y., Schmitt, C., Allemand, Y., Gnemmi, S., Clavel, R., Metrailler, P., & Brodard, R., (2006). The walk trainer: A robotic system for walking rehabilitation. In: *Robotics and Biomimetics* (pp. 1616–1621). ROBIO'06, IEEE International Conference on 2006 Dec. 17. IEEE.

15. Peshkin, M., Brown, D. A., Santos-Munné, J. J., Makhlin, A., Lewis, E., Colgate, J. E., Patton, J., & Schwandt, D., (2005). Kine assists: A robotic overground gait and balance training device. In: *Rehabilitation Robotics* (pp. 241–246). ICORR 2005, 9th International Conference on 2005 Jun 28. IEEE.

16. Zeilig, G., Weingarden, H., Zwecker, M., Dudkiewicz, I., Bloch, A., & Esquenazi, A., (2012). Safety and tolerance of the ReWalk™ exoskeleton suit for ambulation by people with complete spinal cord injury: A pilot study. *The Journal of Spinal Cord Medicine, 35*(2), 96–101.

17. Kawamoto, H., Hayashi, T., Sakurai, T., Eguchi, K., & Sankai, Y., (2009). Development of single leg version of HAL for hemiplegia. In: *Engineering in*

Medicine and Biology Society (pp. 5038–5043). EMBC 2009, Annual International Conference of the IEEE 2009 Sep. 3. IEEE.

18. Quintero, H., Farris, R., Hartigan, C., Clesson, I., & Goldfarb, M., (2011). A powered lower limb orthosis for providing legged mobility in paraplegic individuals. *Topics in Spinal Cord Injury Rehabilitation, 17*(1), 25–33.

19. Ohta, Y., Yano, H., Suzuki, R., Yoshida, M., Kawashima, N., & Nakazawa, K., (2007). A two-degree-of-freedom motor-powered gait orthosis for spinal cord injury patients. *Proceedings of the Institution of Mechanical Engineers, Part H: Journal of Engineering in Medicine, 221*(6), 629–639.

20. Ruthenberg, B. J., Wasylewski, N. A., & Beard, J. E., (1997). An experimental device for investigating the force and power requirements of a powered gait orthosis. *Journal of Rehabilitation Research and Development, 34*(2), 203.

21. Masiero, S., Mastrocostas, M., & Musumeci, A., (2018). Orthoses in older patients. In: *Rehabilitation Medicine for Elderly Patients* (pp. 133–145). Springer, Cham.

22. Butler, P. B., Evans, G. A., Rose, G. K., & Patrick, J. H., (1983). A review of selected knee orthoses. *Rheumatology, 22*(2), 109–20.

23. Draganich, L., Reider, B., Rimington, T., Piotrowski, G., Mallik, K., & Nasson, S., (2006). The effectiveness of self-adjustable custom and off-the-shelf bracing in the treatment of varus gonarthrosis. *JBJS, 88*(12), 2645–52.

24. Hangalur, G., Bakker, R., Tomescu, S., & Chandrashekar, N., (2018). New adjustable unloader knee brace and its effectiveness. *Journal of Medical Devices, 12*(1), 015001.

25. Jonathan, K. P. E., (2009). PhD engineering design review of stance-control knee-ankle-foot orthoses. *Journal of Rehabilitation Research and Development, 46*(2), 257.

26. Berkelman, P., Rossi, P., Lu, T., & Ma, J., (2007). Passive orthosis linkage for locomotor rehabilitation. In: *Rehabilitation Robotics* (pp. 425–431). ICORR 2007, IEEE 10th International Conference on 2007 Jun 13. IEEE.

27. Singh, R., Chaudhary, H., & Singh, A. K., (2017). Defect-free optimal synthesis of crank-rocker linkage using nature-inspired optimization algorithms. *Mechanism and Machine Theory, 116*, 105–122.

28. Singh, R., Chaudhary, H., & Singh, A. K., (2018). A novel gait-based synthesis procedure for the design of 4-bar exoskeleton with natural trajectories. *Journal of Orthopaedic Translation, 12*, 6–15.

29. Singh, R., Chaudhary, H., & Singh, A., (2019). A novel gait-inspired four-bar lower limb exoskeleton to guide the walking movement. *Journal of Mechanics in Medicine and Biology*, 1950020.

30. Kim, J. H., Shim, M., Ahn, D. H., Son, B. J., Kim, S. Y., Kim, D. Y., Baek, Y. S., & Cho, B. K., (2015). Design of a knee exoskeleton using foot pressure and knee torque sensors. *International Journal of Advanced Robotic Systems, 12*(8), 112.

31. Zhang, Y., Arakelian, V., & Le Baron, J. P., (2017). Design of a legged walking robot with adjustable parameters. In: *Advances in Mechanism Design II* (pp. 65–71). Springer International Publishing.

32. McKendry, J., Brown, B., Westervelt, E. R., & Schmiedeler, J. P., (2008). Design and analysis of a class of planar biped robots mechanically coordinated by a single degree of freedom. *Journal of Mechanical Design, 130*(10), 102302.

33. Batayneh, W., Al-Araidah, O., & Malkawi, S., (2013). Biomimetic design of a single DOF Stephenson III leg mechanism. *Mechanical Engineering Research, 3*(2), 43.

34. Sheba, J. K., Elara, M. R., Martínez-García, E., & Tan-Phuc, L., (2017). Synthesizing reconfigurable foot traces using a Klan mechanism. *Robotica, 35*(1), 189–205.

35. Majernik, J., (2013). Normative human gait databases. S*tatistics Research Letters, 2*(3), 69–74.

36. Perry, J., & Davids, J. R., (1992). Gait analysis: Normal and pathological function. *Journal of Pediatric Orthopaedics, 12*(6), 815.

37. Rao, R., (2016). Jaya: A simple and new optimization algorithm for solving constrained and unconstrained optimization problems. *International Journal of Industrial Engineering Computations, 7*(1), 19–34.

Index

Printed and bound by CPI Group (UK) Ltd, Croydon, CR0 4YY

23/10/2024

01777704-0001